Biological and Biogenic Crystallization

Biological and Biogenic Crystallization

Special Issue Editor

Jolanta Prywer

MDPI • Basel • Beijing • Wuhan • Barcelona • Belgrade

MDPI

Special Issue Editor
Jolanta Prywer
Lodz University of Technology
Poland

Editorial Office
MDPI
St. Alban-Anlage 66
4052 Basel, Switzerland

This is a reprint of articles from the Special Issue published online in the open access journal *Crystals* (ISSN 2073-4352) from 2017 to 2018 (available at: https://www.mdpi.com/journal/crystals/special_issues/biogenic_crystallization)

For citation purposes, cite each article independently as indicated on the article page online and as indicated below:

LastName, A.A.; LastName, B.B.; LastName, C.C. Article Title. *Journal Name* **Year**, *Article Number, Page Range.*

ISBN 978-3-03897-521-2 (Pbk)
ISBN 978-3-03897-522-9 (PDF)

Cover image courtesy of Jolanta Prywer.

Contents

About the Special Issue Editor

Jolanta Prywer is currently an associate professor at the Institute of Physics at Lodz University of Technology, Lodz, Poland. She obtained a doctorate in physical sciences awarded by the Faculty of Physics and Chemistry of the University of Lodz, Poland, in 1997. The habilitation procedure was conducted at the Institute of Physics at the University of Silesia in Katowice, Poland, in 2007. She specializes in the analysis and modeling of crystal morphology and phenomena accompanying processes of the crystal growth of various substances. She also deals with processes of biocrystallization in the context of the physiological and pathological mineralization of living organisms. After being promoted to the position of associate professor of Lodz University of Technology (in 2008), she created her own research group called the Biogenic Crystals Research Group, of which she is a leader.

Preface to "Biological and Biogenic Crystallization"

The first biological crystals were grown in the beginning of the 20th century. The first diffraction pattern of biological crystals was done for the enzyme pepsin, which, at the same time, was one of the first enzymes to be crystallized. Soon after that, the tobacco mosaic virus was crystallized. Since that time, biological crystals have become the subjects of intensive research work.

Biogenic crystals are produced by living organisms. They include, for example, calcium oxalate crystals produced in different plant tissues or magnetite crystals forming inside different bacteria and animals or various crystals in the human body appearing in the course of physiological and pathological processes. Biogenic crystals attract a lot of attention because of their fascinating and unique properties.

This book is based on articles submitted for publication in the Special Issue of the Crystals journal, entitled "Biological and Biogenic Crystallization". The intention of this Special Issue was to create an international platform aimed at covering a broad description of results involving the crystallization of biological molecules, including virus and protein crystallization, biogenic crystallization, including physiological and pathological crystallization taking place in living organisms (human beings, animals, plants, bacteria, etc.), and bio-inspired crystallization. Despite many years of research on biological and biogenic crystals, there are still open questions as well as hot and timely topics. This Special Issue contains seven articles that present a cross-section of the current research in the activities in the field of biological and biogenic crystals. The authors of the presented articles prove the vibrant and topical nature of this field. I hope that this Special Issue and this book will serve as a source of inspiration for future investigations and will be useful for scientists and researchers who work on the exploration of biological and biogenic crystals.

I would like to express my deepest gratitude to all authors for their valuable contributions that made this book possible.

Jolanta Prywer
Special Issue Editor

crystals

MDPI

Article

In Situ Random Microseeding and Streak Seeding Used for Growth of Crystals of Cold-Adapted β-D-Galactosidases: Crystal Structure of βDG from *Arthrobacter* sp. 32cB

Maria Rutkiewicz-Krotewicz [1], Agnieszka J. Pietrzyk-Brzezinska [1], Marta Wanarska [2], Hubert Cieslinski [2] and Anna Bujacz [1,*]

[1] Institute of Technical Biochemistry, Faculty of Biotechnology and Food Sciences, Lodz University of Technology, Stefanowskiego 4/10, 90-924 Lodz, Poland; maria.rutkiewicz-krotewicz@dokt.p.lodz.pl (M.R.-K.); agnieszka.pietrzyk-brzezinska@p.lodz.pl (A.J.P.-B.)
[2] Department of Molecular Biotechnology and Microbiology, Faculty of Chemistry, Gdansk University of Technology, Narutowicza 11/12, 80-233 Gdansk, Poland; marta.wanarska@pg.edu.pl (M.W.); hubert.cieslinski@pg.edu.pl (H.C.)
*   Correspondence: anna.bujacz@p.lodz.pl

Received: 29 November 2017; Accepted: 29 December 2017; Published: 1 January 2018

Abstract: There is an increasing demand for cold-adapted enzymes in a wide range of industrial branches. Nevertheless, structural information about them is still scarce. The knowledge of crystal structures is important to understand their mode of action and to design genetically engineered enzymes with enhanced activity. The most difficult task and the limiting step in structural studies of cold-adapted enzymes is their crystallization, which should provide well-diffracting monocrystals. Herein, we present a combination of well-established crystallization methods with new protocols based on crystal seeding that allowed us to obtain well-diffracting crystals of two cold-adapted β-D-galactosidases (βDGs) from *Paracoccus* sp. 32d (*Par*βDG) and from *Arthrobacter* sp. 32cB (*Arth*βDG). Structural studies of both βDGs are important for designing efficient and inexpensive enzymatic tools for lactose removal and synthesis of galacto-oligosaccharides (GOS) and hetero-oligosaccharides (HOS), food additives proved to have a beneficial effect on the human immune system and intestinal flora. We also present the first crystal structure of *Arth*βDG (PDB ID: 6ETZ) determined at 1.9 Å resolution, and compare it to the *Par*βDG structure (PDB ID: 5EUV). In contrast to tetrameric *lacZ* βDG and hexameric βDG from *Arthrobacter* C2-2, both of these βDGs are dimers, unusual for the GH2 family. Additionally, we discuss the various crystallization seeding protocols, which allowed us to obtain *Par*βDG and *Arth*βDG monocrystals suitable for diffraction experiments.

Keywords: β-D-galactosidase; cold-adapted; lactose removal; microseeding; protein crystallization; crystal structure

---

1. Introduction

β-D-Galactosidases (EC 3.2.1.23) are widely used in the food industry as they catalyze the hydrolysis of terminal non-reducing β-D-galactose residue in β-D-galactosides. They are especially relevant in the dairy industry due to their ability to catalyze the hydrolysis of lactose, a natural substrate. Enzymatically hydrolyzed lactose, especially in milk, whey, or whey derivatives, is broadly used due to its higher sweetness, which ameliorates product taste, and to application in specialized food production, for people with lactose malabsorption [1–3]. Administration of products with depleted levels of lactose and other digestible oligosaccharides, disaccharides, monosaccharides, and polyols instead of common food is beneficial in the prevention of irritable bowel syndrome

(IBS) [4–9]. Cold-active β-D-galactosidases (βDGs) have become a focus of attention because of their ability to eliminate lactose from refrigerated milk, convert lactose to glucose and galactose (decreasing its hygroscopicity), and eliminate lactose from dairy industry pollutants associated with environmental problems. Moreover, in contrast to commercially available mesophilic β-D-galactosidase from *Kluyveromyces lactis*, the cold-active enzyme could make it possible to reduce the risk of mesophilic contamination and save energy during the industrial process connected with lactose hydrolysis [10–12].

In addition to hydrolytic activity, some β-D-galactosidases exhibit also a secondary transglycosylation activity, therefore they can be used for the synthesis of oligosaccharides (e.g., GOS and HOS) that are desirable functional food additives [13]. Such an activity is exhibited when there is a high concentration of a substrate and the galactose unit may be transferred onto the substrate, e.g., lactose. The oligosaccharides are built form D-galactose, D-glucose, N-acetylglucosamine, L-fucose, and sialic acid residues linked via O-glycosidic bonds. A vast majority of them carry lactose at their reducing end [14]. GOS are polymers of 2–10 D-galactose units, which are virtually not degraded by human digestive enzymes. Since they reach the colon practically intact and promote the growth of beneficial bacteria (*Bifidobacteria* and *Lactobacilli*), they are classified as prebiotics [15–18]. The importance of GOS as additive to infant formula-milk has been widely discussed as it has been proven not only to promote intestine colonization by beneficial bacteria but also to prevent bacterial adhesion in early stages of infection [14,16,19–24]. Moreover, some oligosaccharides are a rich source of sialic acid (essential for brain development) [25]. GOS and HOS are also valued additives to adult food, as recent research shows that they may increase mineral absorption [26–28], increase the rate of flu recovery, reduce stress-induced gastrointestinal disfunctions [29], as well as prevent cancer formation, benefit lipid metabolism, prevent hepatic encephalopathy, glycemia/insulinemia, and immunomodulation [30,31].

A typical oligomerization state of βDGs from GH2 is tetrameric [11,32] or hexameric [33]. However, large GH2 βDGs were reported to be active as functional dimers, based on biochemical investigations [34,35]. The crystal structure of the first dimeric GH2 βDG was recently published by us [36]; however, that enzyme is much smaller than typical GH2 βDGs (a monomer of only 70 kDa) and exhibits a different shape and orientation of domain 5, called wind-up domain.

Despite extensive efforts and application of different methods for the crystallization of cold adapted proteins, the process is still challenging, as in the Protein Data Bank (PDB) only around 40 crystal structures of cold-adapted enzymes are available, which is a small percentage of 132,000 total structures deposited in the PDB. Whereas the structures of multiple mesophilic βDGs are known, only two structures of cold-adapted βDGs have been previously deposited in the PDB [33,36] and the obtained results show that the investigated enzyme differs in tertiary and quaternary structure from the previously described ones. Here we describe the crystallization methods used to ameliorate and control crystallization of cold adapted *Par*βDG and *Arth*βDG, as well as crystal structure determination of *Arth*βDG, the second cold-active βDG from the GH2 family identified as dimeric up to a year ago.

2. Materials and Methods

2.1. ArthβDG Production

The heterologous expression of the recombinant β-D-galactosidase from *Arthrobacter* sp. 32cB was performed in the *E. coli* LMG 194 cells transformed with pBAD-Bgal 32cB plasmid under the control of P_{BAD} promoter (Table 1). For the production of *Arth*βDG, the *E. coli* cells were grown at 30 °C in Luria-Bertani (LB) medium supplemented with 100 µg/mL ampicillin, until an OD_{600} of 0.5 was reached. Overexpression was induced by addition of 20% L-arabinose solution to the final concentration of 0.02%. The culture was further cultivated for 15 h to OD_{600} of 3.8 ± 0.2 and harvested by centrifugation (6000 × *g*, 15 min, 4 °C) [37].

Table 1. *Arthrobacter* sp. 32cB production information.

Source Organism	*Arthrobacter* sp. 32cB
DNA source	Genomic DNA
Forward primer	F232cBNco (*Nco*I restriction site underlined) TCTA<u>CCATGG</u>CTGTCGAAACACCGTCCGCGCTGGCGGAT
Reverse primer	R32cBHind (*Hind*III restriction site underlined) TGAC<u>AAGCTT</u>CAGCTGCGCACCTTCAGGGTCAGTATGAAG
Cloning vector	pBAD/Myc-His A (Invitrogen, Carlsbad, CA, USA)
Expression vector	pBAD-Bgal 32cB
Expression host	*E. coli* LMG194 (Invitrogen, Carlsbad, CA, USA)

The extraction of intracellular protein was carried out by two separate methods. Method 1: The cells were resuspended in buffer A containing: 20 mM K_2HPO_4/KH_2PO_4 (pH 6.0), 50 mM KCl and the cell suspension was disrupted by sonication on an ice bath using 20 repetitions of 15 s impulses with 60 s pauses to avoid sample overheating. The lysate was clarified by centrifugation at 4 °C for 30 min at 9000 × *g* [37]. Method 2: The cell pellet was ground into a fine powder in a mortar and pestle under liquid nitrogen, with addition of silicone beads. The powder was resuspended in buffer A, and the sample was clarified by centrifugation at 4 °C for 30 min at 9 000 × *g*.

2.2. Purification of ArthβDG

*Arth*βDG was purified by two ion-exchange chromatography steps (weak anion exchanger and strong anion exchanger), followed by a size-exclusion chromatography step. The cell-free supernatant was loaded onto a DEAE (BioRad, Hercules, CA, USA) column equilibrated with buffer A (20 mM K_2HPO_4/KH_2PO_4 (pH 6.0), 50 mM KCl). The recombinant *Arth*βDG was eluted using a linear gradient of potassium chloride (20–1020 mM) in the same buffer. The fractions containing *Arth*βDG were determined and dialyzed against buffer A. In the second step, the protein sample was loaded onto HiPrep Q Sepharose 16/10 column (GE Healthcare, Little Chalfont, UK) equilibrated with buffer A and eluted with a linear gradient of potassium chloride (20–820 mM) in the same buffer. Fractions containing *Arth*βDG were once again determined and dialyzed against buffer C (20 mM K_2HPO_4/KH_2PO_4 (pH 7.5), 150 mM KCl). The concentrated sample was injected onto a Superdex 200 column (GE Healthcare, Little Chalfont, UK), previously equilibrated with buffer C.

The fractions containing *Arth*βDG were identified by SDS-PAGE electrophoresis run at 10% SDS-polyacrylamide gel and by enzymatic activity assay, in parallel. The determination of fractions containing active βDG may be readily validated by enzymatic activity assay: 10% *ortho*-nitrophenyl-β-D-galactopyranoside (ONPG) was added to the sample (1:4 ratio). Sample color change into intense yellow was observed for samples containing βDG. The sample of buffer coming from the chromatography column was changed into 0.05 M HEPES pH 7.0 and the samples were concentrated using 50 kDa cutoff membrane Vivaspin filters (Sartorius, Göttingen, Germany).

2.3. ParβDG and ArthβDG Crystallization

All crystallizations were performed in 24-well plates (Hampton Research, Aliso Viejo, CA, USA) using hanging drop vapor-diffusion method at 18 °C. The 1 μL drop of protein was placed on siliconized glass cover slide, covered with an equal volume of reservoir solution and left to equilibrate against 500 μL of crystallization buffer.

First crystallization conditions for *Par*βDG were found using PEG/Ion Screen™ HR2-126 and Index Screen™ HR2-144 (Hampton Research, Aliso Viejo, CA, USA). Initial optimization of crystallization conditions was performed using varying concentrations of precipitants (PEG MME 2K and ammonium acetate), as well as various pHs. To further improve the crystal morphology, various additives were tried from commercially available Additive Screen (Hampton Research, Aliso Viejo, CA, USA).

Initial crystal screenings used for *Arth*βDG crystallization were: Index Screen™ HR2-144, PEG/Ion Screen™ HR2-126, PEG/Ion2 Screen™ HR2-089 (Hampton Research, Aliso Viejo, CA, USA), and Morpheus® HT-96 (Molecular Dimensions, Suffolk, UK).

2.3.1. Streak Seeding

The seed stock was prepared using the previously obtained hair-type crystals. The crystals were transferred to a tube containing a small volume of reservoir solution (around 20 μL) and the crystals were crushed with a pipette tip. Subsequently, the concentrated seed stock was diluted several times by addition of a larger volume of reservoir solution (~100 μL). The crystallization plate was set up using the previously optimized crystallization conditions. The protein solution was slightly diluted (to 13 mg/mL), premixed with dichloromethane at the final concentration of 0.025% (v/v) and the drops were set in a 1:1 ratio. The cover slides with the drops were sealed and the plate was left for a short pre-equilibration time intact. The hair was run through the seed stock and then through the freshly set drops.

2.3.2. Seed Stock Preparation

The obtained *Arth*βDG crystals were crushed within a crystallization drop with a CrystalProbe (Hampton Research, Aliso Vieja, CA, USA). Next, they were carefully transferred into 50 μL of cool reservoir solution and the solution was vortexed with addition of SeedBead (Hampton Research, Aliso Vieja, CA, USA), kept cool all the time. A series of dilutions of such prepared seed stock was performed and systematic (10×) dilution was used for the rMMS crystallization experiment.

2.3.3. Random Microseed Matrix Screening Crystallization

The procedure rMMS was applied for screening co-crystallization conditions of *Arth*βDG complexes with ligands such as lactose and IPTG. The screening was performed at 18 °C using the sitting drop vapor diffusion technique with an automated sample handling robotic system Oryx4 (Douglas Instruments Ltd., Hungerford, UK) [38]. The drop was composed of: 0.20 μL of protein, 0.07 μL of seed solution and 0.13 μL of reservoir, and placed over 50 μL of reservoir solution. The screens such as PEG/Ion Screen™ HR2-126, PEG/Ion2 Screen™ HR2-098 (Hampton Research, Aliso Vieja, CA, USA), and Morpheus® HT-96 (Molecular Dimensions, Suffolk, UK) were tested for alternative co-crystallization conditions for complexes of *Arth*βDG.

2.4. Data Collection and Processing

Initial X-ray diffraction measurement of the crystals was performed at our home source SuperNova (Rigaku Oxford Diffraction, Tokyo, Japan). High-resolution data were collected using a synchrotron source on beamline BL14.2 at BESSY, Berlin, Germany. For some crystals, 1.8 M sodium malonate, 60% Tacsimate™ (both with appropriate pH) or 50% glycerol solution was used as cryoprotectant during the data collection. Generally cryoprotectants containing only salts of carboxylic acids worked better than those containing glycerol [39]. The diffraction data were processed with XDSapp [40]. The details for the data collection and processing of *Arth*βDG are presented in Table 2.

Table 2. The diffraction data collection and processing statistics for *Arth*βDG crystal PDB ID: 6ETZ.

Diffraction Source	BL 14.2 BESSY, Berlin, Germany
Wavelength (Å)	0.918400
Temperature (K)	100 K
Detector	PILATUS 3S 2M
Crystal-detector distance (mm)	344.48
Rotation range per image (°)	0.1

Table 2. *Cont.*

Diffraction Source	BL 14.2 BESSY, Berlin, Germany
Total rotation range (°)	180
Exposure time per image (s)	0.3
Space group	$P3_121$
a, b, c (Å)	137.78, 137.78, 127.20
α, β, γ (°)	90, 90, 120
Mosaicity (°)	0.077
Resolution range (Å)	46.7–1.8 (1.9–1.8)
Total No. of reflections	1305805
No. of unique reflections	129004
Completeness (%)	99.5 (97.9)
Redundancy	9.98 (9.59)
$I/\sigma(I)$	14.46 (1.5)
R_{meas} (%)	11.3 (136.1)
Overall B factor from Wilson plot (Å^2)	25.6

Values for the outer shell are given in parentheses.

2.5. Structure Solution and Refinement

The Matthews value calculation showed that a monomer of protein is present in the asymmetric unit. The structure of *Arth*βDG was solved by molecular replacement using a monomer of the closest homologue structure (PDB ID: 1YQ2): βDG from *Arthrobacter* C2-2 [33] with the program *PHASER* [41]. The model after rebuilding in *COOT* [42], which was possible due to the significant 2Fo-Fc electron density map for the missing fragments, after first cycle of refinement in *REFMAC5* [43] gave R_{work} and R_{free} values of 19.6% and 23.1%, respectively. That model was further refined in *REFMAC5* using maximum-likelihood targets, including TLS parameters [44] defined for each domain, yielding the final R_{work} and R_{free} of 16.1% and 19.8%, respectively (Table 3).

Table 3. *Arth*βDG crystal structure solution and refinement parameters.

	PDB ID: 6ETZ
Resolution range (Å)	46.73–1.80 (1.84–1.80)
Completeness (%)	99.5
No. of reflections, working set	126888 (8872)
No. of reflections, test set	2100 (146)
Final R_{cryst}	0.161 (0.303)
Final R_{free}	0.198 (0.314)
Cruickshank DPI	0.0913
No. of non-H atoms:	
Protein	7727
Ion	1
Ligand	23
Water	1183
Total	8934
R.m.s. deviations	
Bonds (Å)	0.019
Angles (°)	1.874
Ramachandran plot:	
Most favored (%)	98
Allowed (%)	2

Values for the outer shell are given in parentheses.

3. Results and Discussion

3.1. Crystallization of Cold-Adapted βDGs

3.1.1. Crystallization of Cold-Adapted ParβDG

The crystal structure of *Par*βDG has been already reported in our previous article [36] that focused on its structural analysis (PDB ID: 5EUV). The crystallization process was not discussed there in detail. Therefore, here we describe all steps that were necessary to obtain monocrystals with good diffraction properties.

Since *Par*βDG purification has already been described [36], it will be mentioned only briefly. The protein was expressed in *E. coli* and purified using a two-step protocol employing ion exchange chromatography: the first step was carried out using Fractogel EMD DEAE column (Merck, Darmstadt, Germany) and was followed by a protein separation on a Resource Q column (Merck, Darmstadt, Germany). The active fractions of *Par*βDG were dialyzed against a buffer composed of 0.02 M sodium phosphate, pH 7.3. The protein was concentrated to 15 mg/mL.

The standard crystallization screening performed for *Par*βDG gave clustered hair-like crystals in 24% PEG MME 2K, 0.1 M ammonium acetate, 0.1 M Bis-Tris, pH 6.0 (Figure 1a). After an optimization of the crystallization conditions, a decrease of pH to 5.5 improved slightly the morphology of the *Par*βDG crystals (Figure 1b). Several crystals present in the cluster became thin needles. However, it was still difficult to separate them from the cluster.

Various additives, 96 in total, were tested (Additive Screen, Hampton Research, Aliso Vieja, CA, USA), and the best results were obtained using the previously optimized crystallization conditions and dichloromethane at a final concentration of 0.025% (v/v) as an additive premixed with protein. Some of the crystals still formed hair-type clusters; however, a number of separate needles could be observed in the drops (Figure 1c).

The single needles were obtained by a combination of crystallization with the additive and streak seeding, where the hair was run through a seed stock and then through the freshly set drops. After the second round of seeding, we obtained crystals that grew as separate needles (Figure 1d). The final protein concentration used for setting the drops was 11 mg/mL. The decrease of the protein concentration and introduction of the seeds to the drops enabled significantly improved crystal growth.

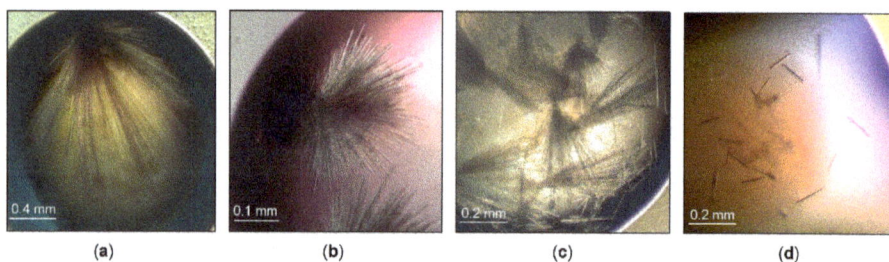

Figure 1. *Par*βDG crystals: (**a**) initial screening (pH 6); (**b**) initial optimization (pH 5.5); (**c**) the optimization with Additive Screen; (**d**) crystals obtained with streak seeding.

3.1.2. Crystallization of Cold-Adapted ArthβDG

*Arth*βDG was produced in *E. coli* as a soluble, intracellular recombinant protein. For its extraction from the cells two methods were used in parallel. Although extraction with mortar and pestle under liquid nitrogen is a time and energy consuming process, it proved to be more beneficial for subsequent protein crystallization than classical sonication. Not only a higher yield of purification was obtained for this extraction method, but the growth of native crystals was more rapid. The first crystals were observed after 3 days, whereas for the protein extracted using sonication the first crystals occurred after 5 days (under corresponding crystallization conditions). Subsequent to extraction, *Arth*βDG was purified using two steps of ion-exchange and the third step was size-exclusion chromatography. The whole purification procedure was conducted at 4 °C, as the protein samples purified at higher temperature (e.g., 18 °C) produced no crystals even under previously determined crystallization conditions. The cold-adapted enzymes exhibit higher propensity towards thermal denaturation [45], the resulting denaturation or unfolding negatively affects subsequent crystallization. The eluted fractions were tested using an enzymatic assay [46], and the ones exhibiting hydrolytic activity versus ONPG were further analyzed by SDS-PAGE electrophoresis. The recombinant *Arth*βDG migrated on an SDS-polyacrylamide gel with a molecular weight of ~110 kDa, which was in agreement with the calculated molecular mass of its monomer based on a cloned construct. The observation of a sharp peak of protein during the last step of purification and the presence of a sole band on an electrophoretic gel proved that the sample was highly purified (Figure 2).

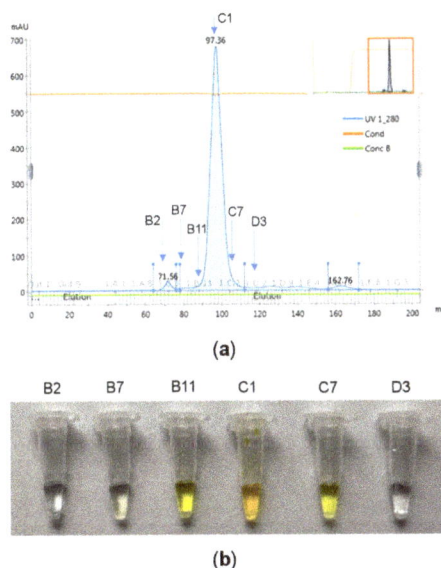

(a)

(b)

Figure 2. *Arth*βDG purification results: (**a**) protein peak purified and concentrated by size-exclusion chromatography, the fractions indicated with arrows were used for enzymatic assay; (**b**) results of the enzymatic assay. The selected protein samples were added to 10% *ortho*-Nitrophenyl-β-D-galactopyranoside (ONPG) solution. The yellow color is produced by ortho-nitrophenol obtained by enzymatic hydrolysis of ONPG, thus indicating fractions with βDG activity (B11, C1, C7), the other tested fractions exhibited no βDG activity (B2, B7, D3); all the fractions from range B11–C7 were combined and used for crystallization experiments.

The initial crystallization screening was performed using the same range of *Arth*βDG concentration (6–12 mg/mL) as we used previously to obtain crystals of cold-adapted aminotransferase from *Psychrobacter* sp. B6 (*Psy*ArAT) [46]. No crystals were observed for a sample concentration below 10 mg/mL. For the protein concentration exceeding 10 mg/mL, a few conditions yielded small crystals, whose size did not exceed 0.02 × 0.02 × 0.02 mm, or microcrystalline precipitate. However, none of them were directly suitable for diffraction experiments. Thus, for crystallization optimization, a sample concentration of 10 mg/mL was used. The optimized crystallization conditions included precipitants such as: sodium malonate, sodium phosphate, potassium phosphate, ammonium sulfate, pentaerythriol etoxylate, L-proline, sodium citrate, PEG 3350, Jeffamine-2001, and Tacsimate™ (Hampton Research, Aliso Vieja, CA, USA). As a result, we obtained larger (up to 0.3 × 0.2 × 0.2 mm)

*Arth*βDG crystals in a tetragonal or bipyramidal form in conditions containing 1.4 M sodium malonate, 1.5% Jeffamine, and 0.1 M HEPES pH 7.0 (Figure 3d), and smaller (up to 0.2 × 0.2 × 0.15 mm) crystals of the same morphology from 35% Tacsimate™ pH 8.0 (Figure 3c). The optimization of precipitating solutions containing PEG 3350 and inorganic salts yielded no monocrystals. No better quality crystals of *Arth*βDG were obtained for crystallization held at 4 °C, regardless of the concept that lowering the experiment temperature, thus thermal energy, should aid crystallization of highly flexible proteins. It might have been caused by its relatively high, as for cold-adapted protein, thermal optimum of 28 °C [37].

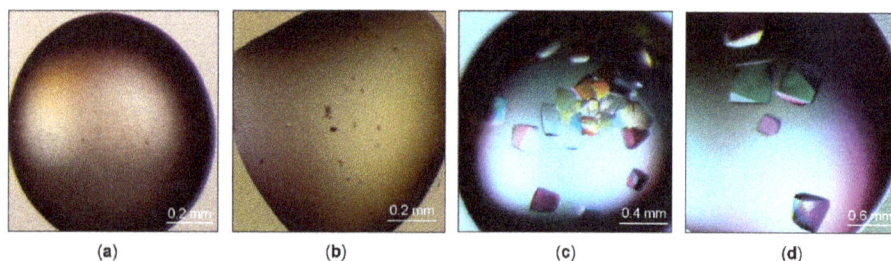

| (a) | (b) | (c) | (d) |

Figure 3. *Arth*βDG crystals. The best results of the initial screening: (**a**) crystallization conditions containing 1.1 M sodium malonate pH 7.0, 0.1 M HEPES pH 7.0, 0.5% Jeffamine® ED-2001 pH 7.0; (**b**) 30% Tacsimate™ pH 7.0; results of the optimization: (**c**) crystallization conditions containing 35% Tacsimate pH 8.0; (**d**) 1.4 M sodium malonate, 0.1 M HEPES pH 7.0, 1.5% Jeffamine® ED-2001 pH 7.0.

The diffraction experiment performed on initial larger crystals (Figure 3b) using a home source SuperNova diffractometer proved that they were of protein; however, their diffraction, due to the small size, was very poor (~8 Å). Further optimization of crystallization conditions and subsequent measurements utilizing synchrotron radiation allowed us to collect complete diffraction datasets with a resolution up to 2.2 Å for large crystals (Figure 3d) (Figure 4a) and up to 1.8 Å for a little smaller (but still big) crystals (Figure 3c) (Figure 4b). Even though the sizes of crystals from 35% Tacsimate™ pH 8.0 were smaller, the resulting diffraction data had higher resolution.

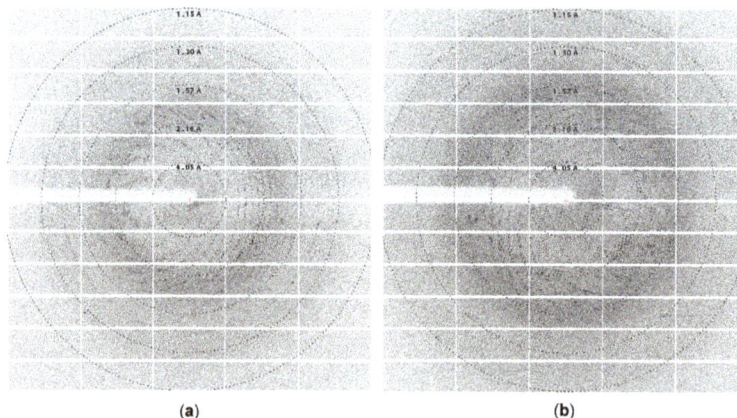

| (a) | (b) |

Figure 4. Diffraction images collected on BL 14.2 line BESSY synchrotron : (**a**) using *Arth*βDG crystal crystallized in the presence of 1.4 M sodium malonate, 1.5% Jeffamine, and 0.1 M HEPES pH 7.0; (**b**) using *Arth*βDG crystal crystallized in the presence of 35% Tacsimate™ pH 8.0.

Since the crystallization conditions included a high concentration of organic acids that might have been preventing ligand binding, the search for alternative crystallization conditions for complexes of *Arth*βDG with isopropyl β-D-1-thiogalactopyranoside (IPTG) and lactose was performed using the random Microseed Matrix Screening (rMMS) procedure. The introduction of seed stock to crystallization drops allowed us to determine multiple crystallization conditions, containing a minimal concentration of Tacsimate™ (introduced with seeds). It is of note that the hits were partially different depending on the used ligand (Figure 5).

Another issue with *Arth*βDG crystals obtained using the classical hanging drop vapor diffusion method was that crystals, which grew in the same drop possessing the same morphology and of a similar size, were randomly diffracting well (~2 Å) or very poorly (~10 Å). To ensure the required quantity of well diffracting crystals, the seeding of the known crystallization conditions was performed. Different seed stock dilutions, 10×, 100×, 1000×, and different protein concentrations, 6 mg/mL, 8 mg/mL, and 10 mg/mL were examined. The use of 8 mg/mL protein concentration and 1000× diluted seed stock yielded formation of ~10 diffracting crystals per drop with average dimensions of 0.25 × 0.2 × 0.2 mm. The adaptation of the seeding procedure for setting up crystallization manually was performed: 6% v/v of cool seed stock was added to the cooled protein solution directly, right before drop setting. The sample was kept on ice for the time of operations. The crystallization was then set up using the standard hanging drop procedure. Introduction of altered crystallization seeding and lowering the protein concentration to 8 mg/mL proved to reproducibly yield diffracting quality crystals. The number of crystals in a drop could be controlled by the use of different dilutions of readily available, pre-prepared and frozen seed stock.

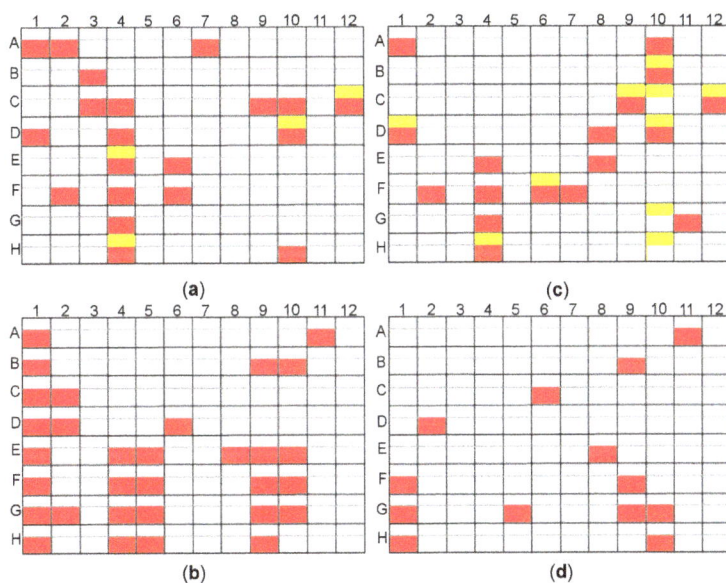

Figure 5. The results of the rMMS experiments for two screen sets depending on the ligand added; yellow–crystals obtained with no seeding (control), red–crystals obtained by seeding; (**a**) PEG/Ion and PEG/Ion2 screen *Arth*βDG co-crystallized with lactose; (**b**) Morpheus and Morpheus II screen *Arth*βDG co-crystallized with isopropyl β-D-1-thiogalactopyranoside (IPTG); (**c**) PEG/Ion and PEG/Ion2 screen *Arth*βDG co-crystallized with IPTG; (**d**) Morpheus and Morpheus II screen *Arth*βDG co-crystallized with lactose. Some of the obtained crystals were big enough for diffraction experiment, e.g., plate (a) A7, plate (b) D8, however most of the obtained crystal hits needed further optimization. Use of microseeding enabled the picking up of a considerable amount of new hits.

3.2. Structure of ArthβDG

3.2.1. Overall Fold

The crystal structure of *ArthβDG* (PDB ID: 6ETZ) revealed that the protein consists of five domains: (Domain 1 (Ser24-Pro224), Domain 2 (Asp225-Gly310), Domain 3 (Ser311-Pro606), Domain 4 (Leu607-Val716), and Domain 5 (Pro717-Ser1010)), with the catalytic one being TIM-barrel, which is typical for glycoside hydrolases. The other four domains, each with an immunoglobulin-like fold, surround Domain 3 and form the outer surface of the functional dimer. Domain 1, Domain 2, and Domain 4 are jelly-roll type barrels. Domain 5 (Pro717-Ser1010) is a large β-sandwich domain (Figure 6). The long linker between Domain 4 and Domain 5, comprising a high number of proline residues, provides some rigidity to this highly solvent exposed fragment. Regardless of low sequence similarity (35%) (Figure S1), *ArthβDG* monomer has an overall fold similar to the characteristics for this group of enzymes *E. coli* lacZ βDG. The intriguing questions, why nature produced such a big multi-domain enzyme to perform a relatively simple reaction, and what is the function of the surrounding domains, still remain open.

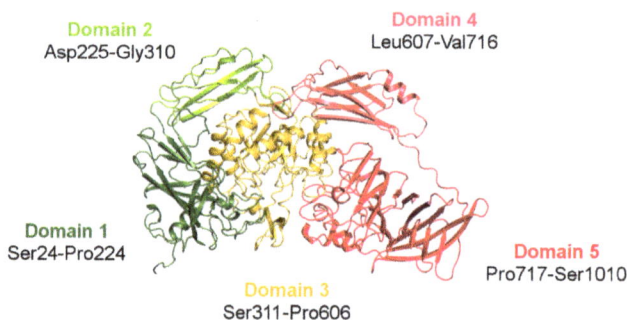

Figure 6. A monomer of *ArthβDG* (PDB ID: 6ETZ) colored by domain. The central catalytic Domain 3 (gold) is typical for the GH2 family TIM-barrel fold. The Domains 1 (dark green) and 5 (red) are responsible for intramolecular contacts forming a biologically active dimer.

3.2.2. Architecture of the Functional Dimer

ArthβDG acts as a head-to-tail dimer, with two independent active sites. Even though only a monomer is present in the asymmetric unit of *ArthβDG* crystal, the functional dimer is mapped by the symmetry operations. The *ArthβDG* dimer interface is made by extensive contacts between Domain 1 interacting with Domain 5 of an adjacent monomer (Figure 7a). *ParβDG* acts also as a functional dimer, which is present in the asymmetric unit of the crystal [36]. The active site of these enzymes is created by residues belonging to both monomers of the dimer. The presence of the dimers in solution was confirmed by gel filtration chromatography (Figure S2), for both enzymes [36]. The *ParβDG* dimer is more elongated and its interface is made by extensive contacts between Domain 3 interacting with Domain 4 of an adjacent monomer and Domains 5 from both monomers (Figure 7b).

The architecture of the functional dimer of *ArthβDG* differs significantly from the functional dimer of *ParβDG* (Figure 7). The buried area of the *ParβDG* dimer is 7180 Å^2, while of the substantially larger *ArthβDG*, 6150 Å^2, the assemblies analysis was performed using a PISA server (Table 4) [47].

Crystal structures of these two enzymes indicate that the biological assembly arrangement is an important factor as to why *ParβDG* exhibits low transglycosylation activity. The cramped vicinity of *ParβDG* active site, and its deep location limits possible galactosyl group acceptors to primarily water molecules. By contrast, the more spacious and easily substrate-accessible active site of *ArthβDG* allows molecules such as galactose, fructose, or salicin, to be acceptors for transfer of the galactosyl group (Figure 8b).

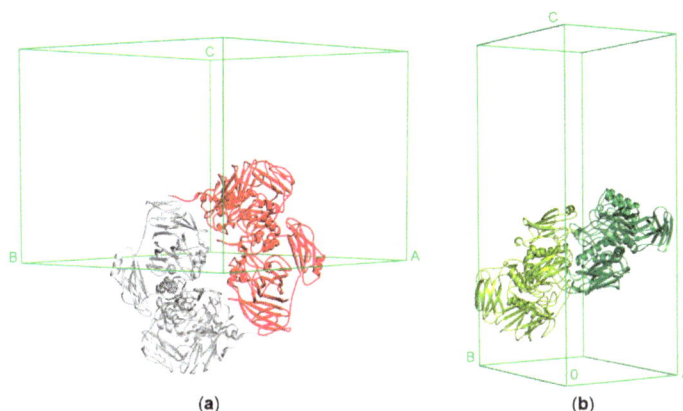

Figure 7. The functional dimers of *Arth*βDG and *Par*βDG in unit cells: (**a**) *Arth*βDG (PDB ID: 6ETZ); (**b**) *Par*βDG (PDB ID: 5EUV).

Table 4. PISA assembly analysis of *Par*βDG and *Arth*βDG.

Protein	Composition	Surface Area [Å²]	Buried Area [Å²]	ΔG^{int} [kcal/mol]	ΔG^{diss} [kcal/mol]
*Par*βDG	dimer	48 760	7 180	−17.5	24.8
*Arth*βDG	dimer	68 700	6 150	−28.9	6.3

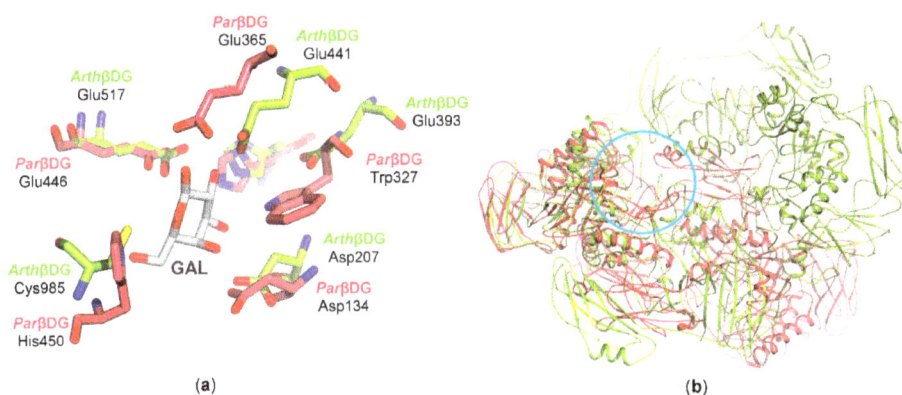

Figure 8. The superposition of *Arth*βDG and *Par*βDG in complex with galactose: (**a**) residues in active sites involved in galactose binding; (**b**) the superposition of biologically active dimers with superposed active sites (marked with a blue circle) of monomers A (*Par*βDG presented in shades of red and *Arth*βDG in shades of green).

3.2.3. Catalytic Center

In both enzymes the catalytic domain is a highly conserved TIM-barrel, which contains eight parallel β-strands and seven α-helices. The catalytic residues were determined to be Glu441, Glu517 in *Arth*βDG and Glu365, Glu446 in *Par*βDG. A superposition of the active sites of *Arth*βDG with the other two cold-adapted βDGs, as well as their superpositions with homologous mesophilic GH2 βDG, revealed that the catalytic center is highly conserved in this group of enzymes.

A superposition of *Arth*βDG with *Par*βDG shows extensive similarity in the region of the catalytic amino acids; however, other parts of the catalytic center exhibit large differences (Figure 8a). Superimposed catalytic centers of both βDGs show that, besides identical catalytic amino acid residues stabilizing galactose binding, His302/His368 and Asp134/Asp207 are also conserved. However, differences in some residues, His450/Cys985 and Trp327/Glu393, result in alteration of active sites shape, volume, and character. The overlay of the biological dimers indicates how the differences in their assemblies influence the substrate selectivity.

The presence of the non-proline *cis*-peptide bonds in the close vicinity of active sites is a very intriguing observation, as such bonds are extremely rare. Two of them are crucial for the creation of an active site, as they constitute its bottom, and therefore regulate its volume. A comparison with other GH2 βDGs revealed that all three observed non-proline *cis*-peptides are present in the same places in this family of enzymes. Moreover, they are common for all glycoside hydrolases that catalyze the reaction with the retaining configuration on anomeric carbon of hydrolyzed β-glycoside.

4. Conclusions

Generally, nucleation is a critical stage in the crystallization of proteins, and for those possessing conformationally labile fragments, as e.g., cold adapted enzymes, it is a bottleneck of crystal structure determination, and it is necessary to implement factors lowering the nucleation barrier.

By implementation of different seeding techniques, such as streak seeding, in situ random microseeding, and streak microseeding, crystallization of two cold-adapted βDGs was successful. Further, the microseeding allowed us to obtain diffraction quality crystals in a routine manner, which significantly simplified further structural studies of these proteins and reduced uncertainties connected with the crystallization stage. Additionally, the used techniques of microseeding allowed growing crystals after nucleation (coming from seeds) without achieving oversaturation.

In the case of cold-adapted proteins, which are more prone to thermal denaturation then mesophilic enzymes, the temperature factor must be taken into account at all stages of protein preparation. In the case of *Arth*βDG, we were able to obtain more protein sample from the same volume of biomass when we used mortar and pestle under liquid nitrogen instead of sonication. The sonication produces a large amount of heat, even though extensive sample cooling was introduced into the protocol, the cold-adapted *Arth*βDG was too sensitive for this method. The next step, chromatographic purification, was even more susceptible to temperature elevation due to its long overall time. We were able to obtain well crystallizing protein samples only when the whole purification procedure was conducted at 4 °C.

The crystal structure of *Arth*βDG, as well as establishing purification and crystallization protocols, gives a good starting point for further crystallographic analysis of this enzyme aiming at its activity engineering [48]. The comparison of its catalytic center, with other βDGs indicates putative residues involved in the transglycosylation reaction.

Supplementary Materials: The following are available online at www.mdpi.com/2073-4352/8/1/13/s1. Figure S1: The sequence alignment of *Arth*βDG and *Par*βDG performed using EMBOSS Needle Pairwise Sequence Alignment. The sequence similarity is 35%, and identity only 17.6% with 47.3% gaps. Catalytic amino acids marked with red boxes; Figure S2: Biochemical oligomerization assays of *Arth*βDG. (a) SDS-PAGE analysis stained with Coomasie Brillant Blue G; lane 1 protein molecular-weight markers (Thermo Fisher Scientific), lane 2 *Arth*βDG; (b) V_e/V_0 versus log MW calibration curve for separation of proteins on a Superdex 200 10/200 GL column with *Arth*βDG marked red (V_e, elution volume; V_0, void volume); (c) chromatographic separation of the fraction containing active *Arth*βDG by gel filtration on a Superdex 200 10/300 GL column.

Acknowledgments: This research was supported by grant 2016/21/B/ST5/00555 (A.B.) from the National Science Centre, Poland. We thank Patrick Shaw Stewart for providing a possibility to test Oryx4 (Douglas Instruments Ltd., East Garston, UK) using our samples. We thank the HZB BESSY Berlin, Germany for providing access to BL 14.2 beamline and MX staff for providing support on beamline. We are grateful to A. Wlodawer, NCI for help with editing the manuscript.

Author Contributions: Maria Rutkiewicz-Krotewicz and Agnieszka J. Pietrzyk-Brzezinska performed crystallization of βDGs; Maria Rutkiewicz-Krotewicz and Anna Bujacz performed synchrotron diffraction data collection, processing, structure solving and carried out structural analysis. Maria Rutkiewicz-Krotewicz purified enzyme, refined the structure and mostly wrote the paper to which Anna Bujacz and Agnieszka J. Pietrzyk-Brzezinska also contributed. Marta Wanarska and Hubert Cieslinski performed enzymes expression in *E. coli* and determined the purification protocol. Anna Bujacz coordinated the project.

Conflicts of Interest: The authors declare no conflict of interest.

References

1. Pritzwald-Stegmann, B.F. Lactose and some of its derivatives. *Int. J. Dairy Technol.* **1986**, *39*, 91–97. [CrossRef]
2. Khan, M.; Husain, Q.; Bushra, R. Immobilization of β-galactosidase on surface modified cobalt/multiwalled carbon nanotube nanocomposite improves enzyme stability and resistance to inhibitor. *Int. J. Biol. Macromol.* **2017**, *105*, 693–701. [CrossRef] [PubMed]
3. Traffano-Schiffo, M.V.; Castro-Giraldez, M.; Fito, P.J.; Santagapita, P.R. Encapsulation of lactase in Ca(II)-alginate beads: Effect of stabilizers and drying methods. *Food Res. Int.* **2017**, *1000*, 296–303. [CrossRef] [PubMed]
4. Borghini, R.; Donato, G.; Alvaro, D.; Picarelli, A. New insights in IBS-like disorders: Pandora's box has been opened; a review. *Gastroenterol. Hepatol. Bed Bench* **2017**, *10*, 79–89. [PubMed]
5. Rossi, M.; Aggio, R.; Staudacher, H.M.; Lomer, M.C.; Lindsay, J.O.; Irving, P.; Probert, C.; Whelan, K. Volatile organic compounds in feces associate with response to dietary intervention in patients with irritable bowel syndrome. *Clin. Gastroenterol. Hepatol.* **2017**, *S1542–S3565*, 31201–31216. [CrossRef] [PubMed]
6. Staudacher, H.M.; Lomer, M.C.E.; Farquharson, F.M.; Louis, P.; Fava, F.; Franciosi, E.; Scholz, M.; Tuohy, K.M.; Lindsay, J.O.; Irving, P.; et al. A diet low in fodmaps reduces symptoms in patients with irritable bowel syndrome and a probiotic restores *Bifidobacterium* species: A randomized controlled trial. *Gastroenterology* **2017**, *153*, 936–947. [CrossRef] [PubMed]
7. Cozma-Petrut, A.; Loghin, F.; Miere, D.; Dumitraşcu, D.L. Diet in irritable bowel syndrome: What to recommend, not what to forbid to patients! *World J. Gastroenterol.* **2017**, *23*, 3771–3783. [CrossRef] [PubMed]
8. Yuce, O.; Kalayci, A.G.; Comba, A.; Eren, E.; Caltepe, G. Lactose and fructose intolerance in Turkish children with chronic abdominal pain. *Indian Pediatr.* **2016**, *53*, 394–407. [CrossRef] [PubMed]
9. Pawłowska, K.; Umławska, W.; Iwańczak, B. Prevalence of lactose malabsorption and lactose intolerance in pediatric patients with selected gastrointestinal diseases. *Adv. Clin. Exp. Med.* **2015**, *24*, 863–871. [CrossRef] [PubMed]
10. Wierzbicka-Wos, A.; Cieslinski, H.; Wanarska, M.; Kozlowska-Tylingo, K.; Hildebrandt, P.; Kur, J. A novel cold-active β-D-galactosidase from the *Paracoccus* sp. 32d-gene cloning, purification and characterization. *Microb. Cell Fact.* **2011**, *10*, 108–119. [CrossRef] [PubMed]
11. Bialkowska, A.M.; Cieslinski, H.; Nowakowska, K.M.; Kur, J.; Turkiewicz, M. A new beta-galactosidase with a low temperature optimum isolated from the Antarctic *Arthrobacter* sp. 20B: Gene cloning, purification and characterization. *Arch. Microbiol.* **2009**, *191*, 825–835. [CrossRef] [PubMed]
12. Cavicchioli, R.; Charlto, T.; Ertan, H.; Mohd Omar, S.; Siddiqui, K.S.; Williams, T.J. Biotechnological uses of enzymes from psychrophiles. *Microb. Biotechnol.* **2011**, *4*, 449–460. [CrossRef] [PubMed]
13. Harju, M. Milk sugars and minerals as ingredients. *Int. J. Dairy Technol.* **2001**, *54*, 61–63. [CrossRef]
14. Kunz, C.; Rudloff, S. Biological functions of oligosaccharides in human milk. *Acta Paediatr.* **1993**, *82*, 903–912. [CrossRef] [PubMed]
15. Boehm, G.; Fanaro, S.; Jelinek, J.; Stahl, B.; Marini, A. Prebiotic concept for infant nutrition. *Acta Paediatr.* **2003**, *92*, 64–67. [CrossRef]
16. Chierici, R.; Fanaro, S.; Saccomandi, D.; Vigi, V. Advances in the modulation of the microbial ecology of the gut in early infancy. *Acta Paediatr.* **2003**, *92*, 56–63. [CrossRef]
17. Bujacz, A.; Jędrzejczak-Krzepkowska, M.; Bielecki, S.; Redzynia, I.; Bujacz, G. Crystal structures of the *apo* form of β-fructofuranosidase from *Bifidobacterium longum* and its complex with fructose. *FEBS J.* **2011**, *278*, 1728–1744. [CrossRef] [PubMed]
18. Salminen, S.; Endo, A.; Scalabrin, D. Early gut colonization with *Lactobacilli* and *Staphylococcus* in infants: The hygiene hypothesis extended. *J. Pediatr. Gastroenterol. Nutr.* **2016**, *62*, 80–86. [CrossRef] [PubMed]

19. Oliveira, D.L.; Wilbey, R.A.; Grandison, A.S.; Roseiro, L.B. Milk oligosaccharides: A review. *Diary Technol.* **2015**, *68*, 305–321. [CrossRef]

20. Lee, L.Y.; Bharani, R.; Biswas, A.; Lee, J.; Tran, L.-A.; Pecquet, S.; Steenhout, P. Normal growth of infants receiving an infant formula containing *Lactobacillus reuteri*, galacto-oligosaccharides, and fructo-oligosaccharide: A randomized controlled trial. *Matern. Health Neonatol. Perinatol.* **2015**, *1*, 9. [CrossRef] [PubMed]

21. Nguyen, T.T.P.; Bhandari, B.; Cichero, J.; Parakash, S. A comprehensive review on in vitro digestion of infant formula. *Food Res. Int.* **2015**, *76*, 373–386. [CrossRef] [PubMed]

22. Musilova, S.; Rada, V.; Vlkova, E.; Bunesova, V. Beneficial effects of human milk oligosaccharides on gut microbiota. *Benef. Microbes* **2014**, *5*, 273–283. [CrossRef] [PubMed]

23. Orrhage, K.; Nord, C.E. Factors controlling the bacterial colonization of the intestine in breastfed infants. *Acta Paediatr.* **1999**, *88*, 47–57. [CrossRef]

24. Li, M.; Monaco, M.H.; Wang, M.; Comstock, S.S.; Kuhlenschmidt, T.B.; Donovan, S.M. Human milk oligosaccharides shorten rotavirus-induced diarrhea and modulate piglet mucosal immunity and colonic microbiota. *ISME J.* **2014**, *8*, 1609–1620. [CrossRef] [PubMed]

25. McVeagh, P.; Miller., J.B. Human milk oligosaccharides: Only the breast. *Acta Paediatr.* **1997**, *33*, 281–286. [CrossRef]

26. Wallace, T.C.; Marzorati, M.; Spence, L.; Weaver, C.M.; Williamson, P.S. New frontiers in fibers: Innovative and emerging research on the gut microbiome and bone health. *J. Am. Coll. Nutr.* **2017**, *36*, 218–222. [CrossRef] [PubMed]

27. Whisner, C.M.; Martin, B.R.; Schoterman, M.H.; Nakatsu, C.H.; McCabe, L.D.; Wastney, M.E.; van den Heuvel, E.G.; Weaver, C.M. Galacto-oligosaccharides increase calcium absorption and gut *Bifidobacteria* in young girls: A double-blind cross-over trial. *Br. J. Nutr.* **2013**, *110*, 1292–1303. [CrossRef] [PubMed]

28. Weaver, C.M.; Martin, B.R.; Nakatsu, C.H.; Armstrong, A.P.; Clavijo, A.; McCabe, L.D.; McCabe, G.P.; Duignan, S.; Schoterman, M.H.; van den Heuvel, E.G. Galactooligosaccharides improve mineral absorption and bone properties in growing rats through gut fermentation. *J. Agric. Food Chem.* **2011**, *59*, 6501–6510. [CrossRef] [PubMed]

29. Hughes, C.; Davoodi-Semiromi, Y.; Colee, J.C.; Culpepper, T.; Dahl, W.J.; Mai, V.; Christman, M.C.; Langkamp-Henken, B. Galactooligosaccharide supplementation reduces stress-induced gastrointestinal dysfunction and days of cold or flu: A randomized, double-blind, controlled trial in healthy university students. *Am. J. Clin. Nutr.* **2011**, *93*, 1305–1311. [CrossRef] [PubMed]

30. Mussatto, S.I.; Mancilha, I.M. Non-digestible oligosaccharides: A review. *Carbohydr. Polym.* **2007**, *68*, 587–597. [CrossRef]

31. Swennen, K.; Courtin, C.M.; Delcour, J.A. Non-digestible oligosaccharides with prebiotic properties. *Crit. Rev.. Food Sci. Nutr.* **2006**, 459–471. [CrossRef] [PubMed]

32. Juers, D.H.; Heightman, T.D.; Vasella, A.; McCarter, J.D.; Mackenzie, L.; Withers, S.G.; Matthews, B.W. A structural view of the action of *Escherichia coli* (lacZ) β-galactosidase. *Biochemistry* **2001**, *40*, 14781–14794. [CrossRef] [PubMed]

33. Skalova, T.; Dohnalek, J.; Spiwok, V.; Lipovova, P.; Vondrackova, E.; Petrokova, H.; Duskova, J.; Strnad, H.; Kralova, B.; Hasek, J. Cold-active beta-galactosidase from *Arthrobacter* sp. C2-2 forms compact 660 kDa hexamers: Crystal structure at 1.9 Å resolution. *J. Mol. Biol.* **2005**, *353*, 282–294. [CrossRef] [PubMed]

34. Faning, S.; Leahy, M.; Sheehan, M. Nucleotide and deduced amino acid sequences of *Rhizobium meliloti* 102F34 *lacZ* gene: Comparison with prokaryotic beta-galactosidases and human beta-glucuronidase. *Gene* **1994**, *141*, 91–96. [CrossRef]

35. Burchhardt, G.; Bahl, H. Cloning and analysis of the beta-galactosidase-encoding gene from *Clostridium thermosulfurogenes* EM1. *Gene* **1991**, *106*, 13–19. [CrossRef]

36. Rutkiewicz-Krotewicz, M.; Pietrzyk-Brzezinska, A.J.; Sekula, B.; Cieslinski, H.; Wierzbicka-Wos, A.; Kur, J.; Bujacz, A. Structural studies of a cold-adapted dimeric β-D-galactosidase from *Paracoccus* sp. 32d. *Acta Crystallogr. D* **2016**, *72*, 1049–1061. [CrossRef] [PubMed]

37. Pawlak-Szukalska, A.; Wanarska, M.; Popinigis, A.T.; Kur, J. A novel cold-active β-D-galactosidase with transglycosylation activity from the Antarctic *Arthrobacter* sp. 32cB-gene cloning, purification and characterization. *Process. Biochem.* **2014**, *49*, 2122–2133. [CrossRef]

38. Shaw Stewart, P.D.; Kolek, S.A.; Briggs, R.A.; Chayen, N.E.; Baldock, P.F.M. Random microseeding: A theoretical and practical exploration of seed stability and seeding techniques for successful protein crystallization. *Cryst. Growth Des.* **2011**, *11*, 3432–3441. [CrossRef]
39. Bujacz, G.; Wrzesniewska, B.; Bujacz, A. Cryoprotection properties of salts of organic acids: A case study for a tetragonal crystal of hew lysozyme. *Acta Crystallogr. D* **2010**, *66*, 789–796. [CrossRef] [PubMed]
40. Sparta, K.M.; Krug, M.; Heinemann, U.; Mueller, U.; Weiss, M.S. Xdsapp2.0. *J. Appl. Crystallogr.* **2016**, *49*, 1085–1092. [CrossRef]
41. McCoy, A.J.; Grosse-Kunstleve, R.W.; Adams, P.D.; Winn, M.D.; Storoni, L.C.; Read, R.J. Phaser crystallographic software. *J. Appl. Cryst.* **2007**, *40*, 658–674. [CrossRef] [PubMed]
42. Emsley, P.; Cowtan, K. Coot: Model-building tools for molecular graphics. *Acta Crystallogr. D* **2004**, *60*, 2126–2132. [CrossRef] [PubMed]
43. Murshudov, G.N.; Vagin, A.A.; Dodson, E.J. Refinement of macromolecular structures by the maximum-likelihood method. *Acta Crystallogr. D* **1997**, *53*, 240–255. [CrossRef] [PubMed]
44. Winn, M.D.; Isupov, M.N.; Murshudov, G.N. Use of TLS parameters to model anisotropic displacements in macromolecular refinement. *Acta Crystallogr. D* **2001**, *57*, 122–133. [CrossRef] [PubMed]
45. Gerday, C. Psychrophily and catalysis. *Biology* **2013**, *2*, 719–741. [CrossRef] [PubMed]
46. Bujacz, A.; Rutkiewicz-Krotewicz, M.; Nowakowska-Sapota, K.; Turkiewicz, M. Crystal structure and enzymatic properties of a broad substrate-specificity psychrophilic aminotransferase from the Antarcticsoil bacterium *Psychrobacter* sp. B6. *Acta Crystallogr. D* **2015**, *71*, 632–645. [CrossRef] [PubMed]
47. Krissnel, E.; Henrick, K. Interference of macromolecular assemblies from crystalline state. *J. Mol. Biol.* **2007**, *372*, 774–797. [CrossRef] [PubMed]
48. Talens-Perales, D.; Polaina, J.; Marín-Navarro, J. Enzyme Engineering for Oligosaccharide Biosynthesis. In *Frontier Discoveries and Innovations in Interdisciplinary Microbiology*; Springer: New Delhi, India, 2016; Chapter 2; pp. 9–31.

crystals

MDPI

Article

Formation Mechanism of CaCO$_3$ Spherulites in the Myostracum Layer of Limpet Shells

Shitao Wu, Chang-Yang Chiang and Wuzong Zhou *

EaStCHEM, School of Chemistry, University of St Andrews, Fife KY16 9ST, UK;
sw236@st-andrews.ac.uk (S.W.); cyc3@st-andrews.ac.uk (C.-Y.C.)
* Correspondence: wzhou@st-andrews.ac.uk; Tel.: +44-1334-467276

Academic Editor: Helmut Cölfen
Received: 14 August 2017; Accepted: 16 October 2017; Published: 23 October 2017

Abstract: CaCO$_3$ spherulites were found in the myostracum layer of common limpet shells collected from East Sands, St Andrews, Scotland. Their microstructures were revealed by using powder X-ray diffraction, scanning electron microscopy, high-resolution transmission electron microscopy, and energy dispersive X-ray microanalysis. The formation mechanisms of these spherulites and their morphology evolution were postulated. It was proposed that spherical particles of an inorganic and biological composite formed first. In the centre of each spherical particle a double-layer disk of vaterite crystal sandwiching a biological sheet developed. The disk crystal supplies a relatively strong mirror symmetric dipole field, guiding the orientations of the nanocrystallites and the arrangement of mesorods and, therefore, determining the final morphology of the spherulite.

Keywords: CaCO$_3$; spherulite; biomineralization; limpet shells; electron microscopy; dipole field

1. Introduction

Calcium carbonate is largely found in natural creatures, especially marine organisms, such as echinoderms, abalone, and sea molluscs [1,2]. These parts of inorganic-organic composites often function as body supports, e.g., exoskeletons, to protect the creatures from injury, or form shelters to offer themselves a safe living environment. Unlike relatively regular artificial crystals, CaCO$_3$ crystals in biological systems present very complicated morphologies, e.g., the aragonite nanoplates in nacre [3], the complex structure of coccolithophores [4], calcite spines of sea urchins [5], etc., which seem to be perfectly pre-designed and are rarely reproduced by using any advanced biomimetic methods in the laboratory. It is, therefore, of great interest to reveal detailed processes of crystal growth and to understand crucial factors which control the crystal growth in natural creatures.

Crystalline CaCO$_3$ exists as one of three principal anhydrous polymorphs, i.e., calcite (rhombohedral), aragonite (orthorhombic), and vaterite (hexagonal). According to the phase diagram of pure calcium carbonate, calcite has been proved to be the most thermodynamically stable phase, and vaterite, the least [6]. Vaterite would transform into a more stable polymorph upon contact with water [7], e.g., converting to calcite at low temperature and to aragonite at high temperature. However, in some special conditions, the stabilities of these three polymorphs can alternate. For example, Mg substitution in calcite may greatly reduce its stability, leading to a transformation to aragonite [8].

In nature creatures, CaCO$_3$ microcrystals are commonly embedded in a biological matrix and both aragonite and vaterite can be the stable phases. For example, aragonite is included in many marine biominerals, such as nacreous tablets [9–12], urchin spines [5,13], and crossed lamellae in seashells [14]. Metastable vaterite can also exist in the natural productions and biomineralization. According to the research of Soldati et al. [15] and Wehrmeister et al. [16], both vaterite and aragonite were detected in freshwater cultured pearls, and vaterite was observed to exist in an aragonite environment. In Ascidiacea, known as sea squirts, vaterite also exists in form of spicules [17].

In addition to these crystalline phases, in many biomineralization processes, amorphous calcium carbonate (ACC), as a precursor of crystalline $CaCO_3$, plays an important role in the formation of crystalline phases and crystal morphologies. In a simple environment, ACC can transform to aragonite at high temperature (over 40 °C) and to calcite at low temperature [18,19].

Spherulite is probably one of the most interesting morphologies found in natural creatures or biominerals, because their nature of the inorganic and biological composite and their radial crystal orientations. Spherulites were found in a wide range of materials, in addition to the common carbonates, for instance, polymers [20], metals [21], and organic molecules [22]. Spherulites have also attracted increasing attention from geologists [23–25], because carbonate spherulites have high porosity and permeability and are excellent reservoirs in pre-salt fields. In the last decades, several key factors of forming spherulites have been proposed, including Brownian motion, phase field effect, and dipole field interaction.

Spherulites can form via aggregation of tiny nanocrystallites (<10 nm) in Brownian motion [26]. However, the detailed microstructures and complicated morphologies of spherulites cannot be explained simply using Brownain motion. The phase field effect was believed to be a more convincing explanation of the spherulitic growth [27]. According to the previous research, the spherulites were divided into two categories: (1) spherulites grow in a radial manner from a nucleus and more branches intermittently filled the space during the growth; and (2) threadlike particles grow first and new grains formed at the growth front. By splaying out the branching on both ends, a spherical morphology was achieved. In this process, two characteristic "eyes" were developed on each side of the core region [27]. The phase field, as is claimed to be the driving force of the spherulite formation, was based on the "diffusional instabilities". The model indicated a competition between the ordering effect of discrete local crystallographic symmetries and the randomization of the local crystallographic orientation with growth front nucleation. In 2000, Banfield et al. [28] revealed a three-dimensional rotation of nanoparticles in their HRTEM study of natural iron oxyhydroxide biomineralization products. Neighbouring nanoparticles could aggregate and rotate to adopt parallel orientations.

According to many published papers, some elements, such as magnesium, can not only affect the stability of the crystalline $CaCO_3$, but also play important roles in the formation of spherulitic morphology. Tracy et al. reported that co-existence of Mg^{2+} and SO_4^{2-} ions allowed a formation of spherulites of $CaCO_3$ [29,30]. ACC was also found to be important to the formation of novel morphologies, in addition to working as a transient precursor for crystalline phases of $CaCO_3$ [31]. It is quite interesting to see that, in biological systems, some organisms can produce unusually stable hydrated ACC, as reported by Addadi et al. [32]. The hydration of ACC would inevitably affected by pH in the solution [33]. The crystallization process of ACC in biomineralization was also investigated by using an in situ small- and wide-angle X-ray scattering (SAXS/WAXS) method, and a multi-step evolution was discovered, i.e., formation in order of hydrated and disordered ACC, more ordered and dehydrated ACC, ACC dissolution and spherulitic growth of vaterite, and Ostwald ripening of the vaterite [34].

The addition of organic compounds in a synthetic system can enhance the formation of spherulitic morphology of $CaCO_3$. For example, hollow spherulites of calcite crystals would precipitate in the presence of alginates [35]. Biological matter is even better. Rodriguez-Navarro et al. believed that bacterial production of CO_2 and NH_3, and the transformation of the latter to $OH-$, led to the growth of vaterite crystallites on bacterial cells, or the so-called extracellular polymeric substance (EPS), to form spherulites [36]. Greer et al. [37] also observed such a phenomenon in a synthesis of biomimetic vateritic hexagonal prisms with gelatine, and further found that the electrical dipole field interaction between the crystallites was the driving force for aggregation and self-orientation. In other words, the nanocrystallites were regarded as dipoles and, because of the interaction with the central dipole field, the nanocrystallites embedded in a soft matter matrix could rotate and self-arrange in an unusual way.

There is another common argument of whether the spherulites of CaCO₃ grow directly from solution by central multidirectional growth or by aggregation of nano-sized precursor crystals. Andreassen and co-workers made a significant effort and found that the former is more likely the true mechanism [38,39]. This conclusion is probably correct for some spherulites, such as the particles constructed by a single layer of radiating nanorods [40]. However, it may face a challenge in explaining the formation of spherulites with multiple layers of short nanorods. Further evidence is needed from HRTEM images of the specimens at intermediate stages of growth.

In the present work, spherulites found in limpet shells are investigated chiefly by using electron microscopy. Some interesting microstructures have been observed, e.g., a core rich in biological matter, multilayer short nanorods, possible evolution of the microstructures, etc. A dipole field-driven formation mechanism of the hierarchical structure of these calcium carbonate mesospherulites is proposed. Spherulites formed in a natural environment and synthesized in the laboratory, as presented in a previously-published report [37], are compared in order to shed light to the future biomimetic synthesis of crystals with various morphologies.

2. Results

Live limpet samples were collected at East Sands, St Andrews, Scotland (Figure 1a). According to SEM observations, a limpet shell consists of six layers (M − 2, M − 1, M, M + 1, M + 2, M + 3, from inner to outer sides), as illustrated in Figure 1c. Spherulites about 100 μm in diameter were only found in the M layer (abbreviation of myostracum) (Figure 1b). There is an equatorial notch at the centre of each spherulite, dividing the whole particle into two hemispheres, which was the first sign that aroused our curiosity.

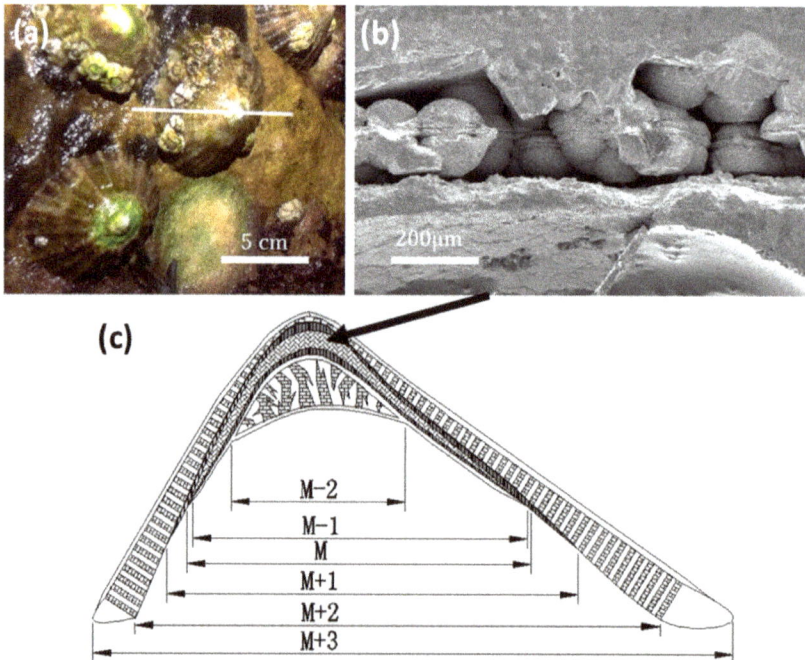

Figure 1. (a) Live limpets on site. (b) SEM image of some spherulites in the myostracum layer. (c) Schematic drawing of a cross-section by cutting from the apex of a limpet shell. The cutting position is indicated by a white line in (a). Six layers are marked. The arrow points to the myostracum layer where the spherulites were found.

SEM images of whole spherulites show a rough surface (Figure 2a). Examinations of the exterior and near surface area of the spherulites reveal that the outer layer of the particles is formed by radially-located mesorods, which have a shape of a hexagonal prism with about 1.5–5 µm in width and 20 µm in length, sticking out from the centre of the particles (Figure 2b). On the side surface of the hexagonal prisms, multiple nanosheets grow from the bottom as coating skins, indicating that the starting point of the growth is at the bottom ends of the mesorods and the mesorods have a high crystallinity. Similarly, aragonite crystal mesorods were synthesized by Zhou et al. and Nan et al. [41,42].

The equatorial notches on the spherulites are not only a surface structural feature, but evidence of cracks cross to the particle cores. Figure 2c shows a SEM image of a cross-section of spherulite, revealing the microstructure in the central area of the particle. It is noted that the crack starts from a small disk (~4 µm in diameter) at the centre of a core, which is about 15 µm in diameter. The spherical core is separated into two hemispheres by the smaller disk, which seems to have a sandwich shape (see inset of Figure 2c).

Figure 2d shows a SEM cross-section image of another spherulite, which is also supposed to be at an early stage of growth because of the small size and the relatively poor ordering of the nanoparticles. In particular, radiating mesorods have not formed in the surrounding areas, marked D. It can be seen that the profile view of the central disk shows a multi-layer structure, i.e., a plate sandwiched by two layers (B). The central plate (elliptical shape in a profile view marked by A), about 1.1 µm in thickness and 4 µm in diameter, likely contains only biological substances, rather than Ca-containing inorganic molecules, as judged from the dark image contrast. A similar area was examined by using EDX to be Ca poor and carbon rich (see Figure 6 below). The two adjacent layers marked by B, 2.5 µm in thickness and 15 µm in diameter, show possible $CaCO_3$ crystals. Outside of layer B formed another layer marked by C, in which radially-ordered traces can be seen, implying that nanocrystallites in this layer have self-orientated, probably guided by the central disk. Outside of layer C, in the regions marked by D, there are many parallel, short, needle-like particles lying along a direction which seems to have no special relation with the mesorods in the central area. This area seems to be less affected by the central disk.

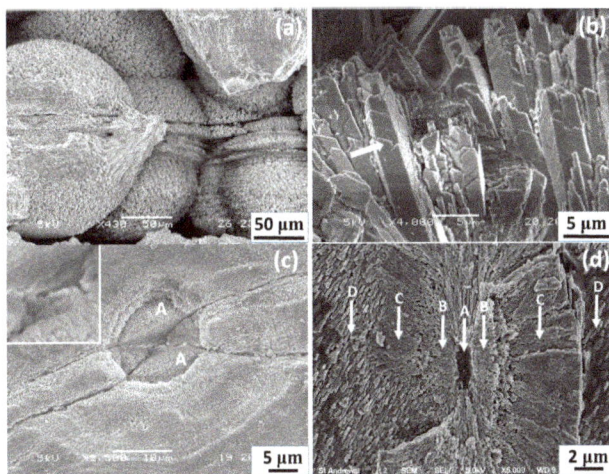

Figure 2. SEM images of some spherulites at high magnifications. (**a**) Whole spherulites with a rough surface. (**b**) Morphology of mesorods in the outer layer of a spherulite. The arrow indicates a skin of a mesorod. (**c**) Cross-section of a central area of spherulite. The inset is an enlarged image showing the double-layer disk in the centre. The two hemispheres are marked by "A". (**d**) Higher magnification image of the sandwich disk (B-A-B) of a spherulite at an early stage. The different layers from the centre are marked from A to D.

Although the crystal structure of individual crystallites can be determined by using selected area electron diffraction (SAED) patterns and high-resolution transmission electron microscopy (HRTEM) images, powder X-ray diffraction (XRD) is used to determine all possible crystalline phases in the whole shell. A limpet shell was ground into powder for XRD analysis. According to the XRD result shown in Figure 3, all three common phases of calcium carbonate, i.e., calcite, aragonite, and vaterite, are detected. The major crystalline phases are calcite and aragonite. The calcite phase has a rhombohedral structure with a = 4.98 Å, c = 17.07 Å, space group R-3c (JCPDS card no. 01-080-9776) and the aragonite phase is orthorhombic with a = 4.97 Å, b = 7.96 Å, c = 5.75 Å, space group Pmcn (JCPDS card no. 01-075-9984). There are some small peaks confirming the existence of hexagonal vaterite as a minor phase in the limpet shell with the unit cell parameters a = 4.13 Å, c = 8.49 Å, space group P63/mmc (JCPDS card no. 24-0030 [43]).

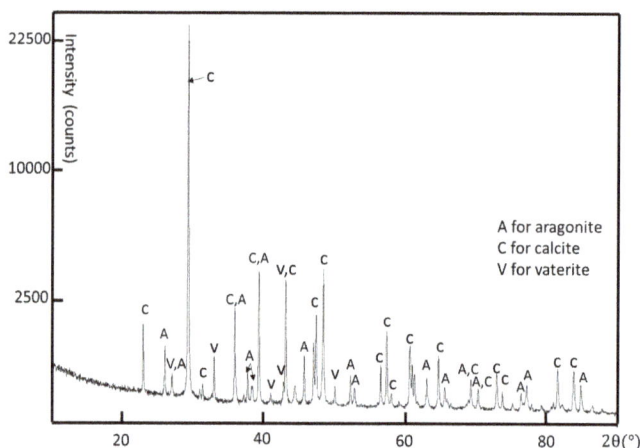

Figure 3. Powder XRD pattern of a crushed limpet shell, showing the presence of calcite (C), aragonite (A), and vaterite (V) $CaCO_3$.

To explore the structure of the mesorods in the spherulites, a focused ion beam (FIB) technique was used to prepare samples of mesorods for the HRTEM study. A slice of a mesorod 200 nm in thickness and 15 µm in length along the long axis was cut out. Differing from the expected high crystallinity and uniform thickness, low-magnification transmission electron microscopy (TEM) images of the mesorod show a flocculent image contrast pattern, implying a large number of defects (Inset of Figure 4a). HRTEM images reveal that the surface area of the mesorod has a poor crystallinity, containing many nanocrystallites embedded in an amorphous matrix (Figure 4a). These nanocrystallites are randomly orientated.

In an inner area shown in Figure 4b, nanocrystallites are also recognized. However, they are all self-orientated and connected. The corresponding FFT pattern looks like a type of single crystal. Moving towards the centre of the mesorod, high crystallinity regions can be found, as shown in Figure 4c. The crystal orientations of the regions in Figure 4b,c are the same. A SAED pattern, recorded from a large area of this sample covering both polycrystalline region and high crystallinity region, is similar to the FFT patterns (Figure S1 in Supplementary Information). The two principal d-spacings are measured from the HRTEM images in Figure 4b,c, d_A = 5.73 Å and d_B = 7.95 Å, with an interplane angle of 90°, which can be indexed to the (001) and (010) planes of the aragonite structure. It is also confirmed that the longitudinal axis of the mesorod is parallel to the [001] zone axis of aragonite. According to the space group Pmcn of aragonite, the (001) and (010) diffraction spots should be systematically absent. Their appearance can be attributed to multiple scattering. However, lattice distortion lowering the symmetry of the aragonite structure cannot be ruled out.

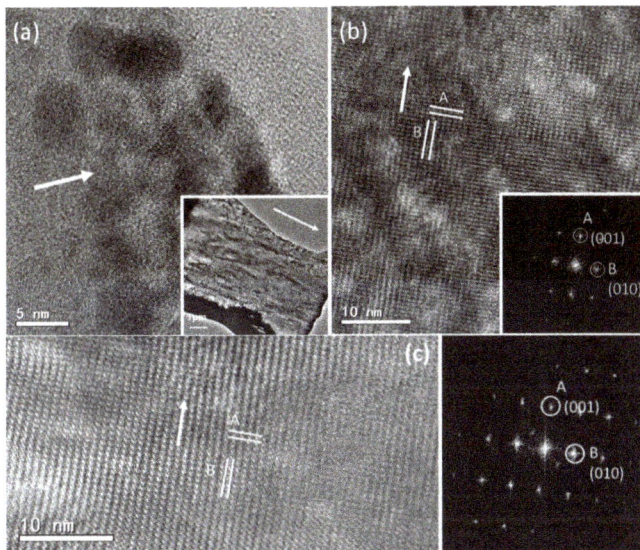

Figure 4. HRTEM images of a mesorod in a spherulite from edge to centre, (**a**) polycrystalline edge with randomly-orientated nanocrystallites, (**b**) an area with all nanocrystallites self-orientated and connected, and (**c**) a single crystalline region. The arrows indicate the long axis of the mesorod. The inset of (**a**) shows a TEM image of the mesorod and the insets in (**b**) and (**c**) are the corresponding FFT patterns. The diffraction spots are indexed to the aragonite structure.

Some spherulite particles (as shown in Figure 5) are elliptical with a relatively smooth surface (denser), but without a visible notch. To gain more information of their inner structures, it is essential to have an insight of the cross-section. The particles were cut into halves and the newly-exposed surfaces were polished. A cross-section SEM image of a typical spherulite is shown in Figure 5a. The spherulite is not a perfect spherical particle. The notch in the middle can still be recognized, although it is difficult to see from outside. The notch extends through the spherulite and separates the spherulite into symmetrical hemispheres. Along the notch, the particle becomes "thin" with a waist and looks like a short dumbbell. On the other hand, the spherulite shown in Figure 5b has no obvious notch at all, although a similar dumbbell-like core is maintained.

The hierarchical structure of the spherulites can be described as multi-layer spherical particles and over 10 distinguished layers can be detected in each spherulite (Figure 5a,b). One micrometre-wide low-density gaps are presented in between the layers (Figure 5c). The layer thickness gets smaller when the layer is closer to the core.

These layers are formed with mesorods, which are all in a radial arrangement. The diameter of the mesorods also decreases when they are closer to the core (Figure 5c). The enlarged SEM image of individual mesorods show that the mesorods in inner layers do not have a regular hexagonal morphology, as we observed from the mesorods in an outer layer (see Figure 2b). The relatively thin mesorods seem to have a polymer-like coating layer, referring to the morphology, which likely consists of biological substances. There is also a gap between the mesorods, making the layer highly porous.

The core structure must be crucial in constructing the microstructures of spherulites. Most particles have a double-hemisphere core. The first image in Figure 6 is a SEM image of a half spherulite. The image contrast in the two hemispheres separated by a double-layer disk particle is significantly darker than other parts, implying that these areas contain light elements or have a low density. Elemental mapping shows that the carbon map is brighter than calcium and oxygen, and the Ca and O concentrations are very low in these double-hemispherical areas in the core, further indicating that these areas contain less $CaCO_3$, but rich in organic materials (Figure 6). This structure is quite different from that in Figure 2d, where only one biomass rich area is present, i.e., the central plate inside the double-layer disk.

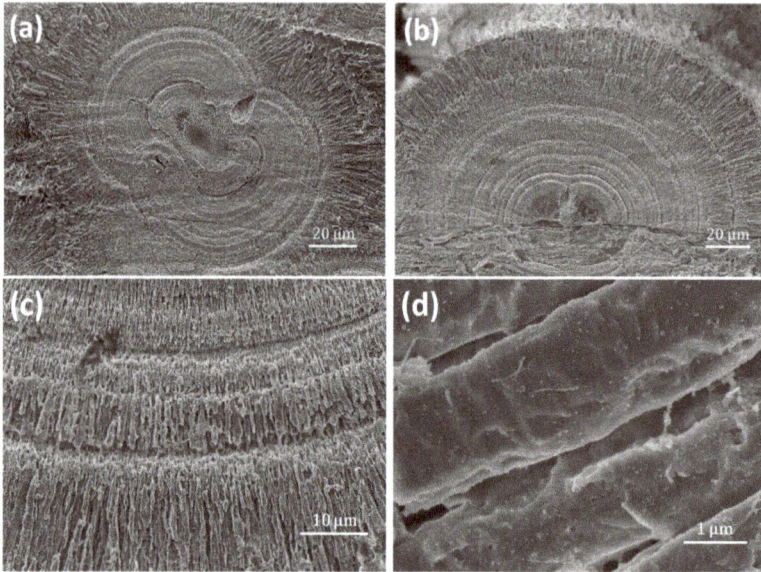

Figure 5. SEM images of cross-sections of spherulites, showing (**a**) an entire spherulite with the notch about to disappear, (**b**) half-spherulite without a notch, (**c**) porous mesorod layers in a spherulite, and (**d**) individual mesorods in spherulites.

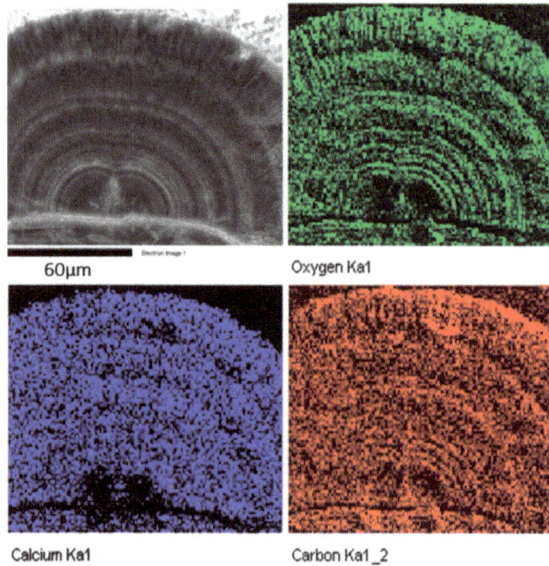

Figure 6. Cross-section SEM image of a half spherulite, and the corresponding EDX elemental mapping showing distributions of O, Ca, and C.

It is believed that the core particles control the morphology and hierarchical structures of the spherulites. To fully understand the formation of the cores, it is useful to look at the cores in early ages when they are small and have a low crystallinity. The particles shown in Figure 7 are probably some examples of these core particles and surroundings. Such an assumption has to be made since we are unable to observe a real growth of a spherulite with time. Assuming Figure 7a shows a SEM image of the cross-section of an early stage spherical particle, it is noticed that before the formation of the mesorods, a small core of about 5 μm in diameter is formed. The core is separated from the surroundings, where many nanocrystallites are detected.

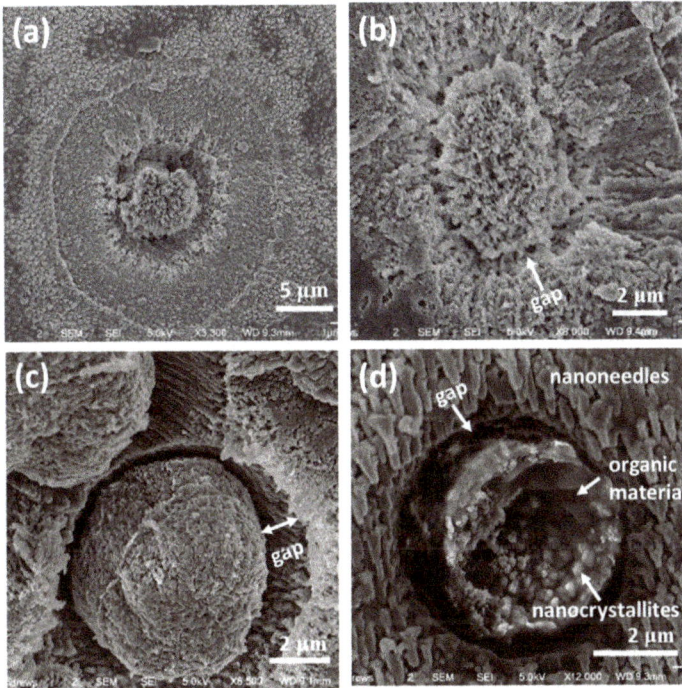

Figure 7. FEG-SEM images of the sea urchin-like particles in the central areas of some early stage spherulites. (**a**) Cross-section of the whole particle when nanorods/mesorods have not formed. (**b**) Another core particle with a low density in the early stage of growth. (**c**) A larger, high-density spherical core separated from the surroundings. (**d**) A core particle with many nanocrystallites embedded in biological substances and some short nanoneedle particles, lying along the same orientation, are developed in the surroundings.

Figure 7b shows probably another core particle with a low density. The core with a rough surface looks like a sea urchin. The core particle in Figure 7c is larger, and denser. The surface is smoother than the cores in Figure 7a,b, and the gap of about 1 μm between the core and the surroundings is obvious. In Figure 7d, the broken core has similar features. However, it can be seen that many nanoparticles are embedded in a matrix, which shows a gel-like dark and smooth contrast. On the other hand, the surroundings is consist of many parallel short nanoneedle particles (~1 μm long and 0.4 μm wide).

Some even smaller spherical particles were also detected by TEM and their crystallinity was examined by HRTEM. These samples were quite unstable under electron beam irradiation. This is understandable because the crystallites were embedded in an organic matrix. However, if proper specimen treatments and microscopic operation for a low-dose irradiation are performed, HRTEM images of crystal structures of many beam-sensitive samples can still be possible [44]. The TEM image in Figure 8a shows some spherical particles (50 to 200 nm in diameter) found in the M layer in a limpet shell. A closer examination of the particles revealed that most parts of these particles are amorphous (see the inset of Figure 8a), which could be a mixture of ACC and biological components. However, many randomly-oriented nanocrystallites in the amorphous matrix are also detected in some other regions. Figure 8b shows lattice fringes of some nanocrystallites. Two kinds of *d*-spacings of these nanocrystallites, 2.133 Å and 2.733 Å, were measured and can be indexed to the (004) and (102) planes of vaterite. Similar results were reported in biominerals of earthworm's calciferous gland by Gago-Duport et al. [45], where vaterite nanocrystallites were developed in ACC particles.

The short nanoneedle particles in the surroundings near the cores (see Figure 7d) are monocrystalline with only a very thin amorphous coating. Figure 9 is an HRTEM image of two cross-linked nanoneedles and the corresponding FFT pattern. The measured *d*-spacings, $d_A = 3.67$ Å and $d_B = 3.62$ Å with an interplane angle of 60°, can be indexed to the (100) and (010) planes of the hexagonal vaterite structure.

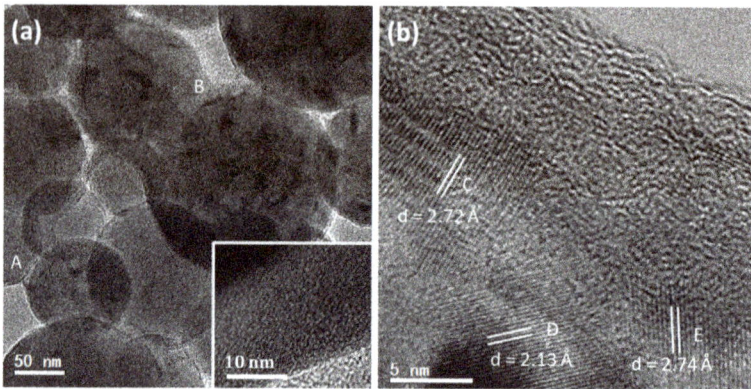

Figure 8. (**a**) TEM image of small spherical particles found in the M layer in a limpet shell. The inset is a HRTEM image recorded from area A. (**b**) A HRTEM image from area B in (**a**). Some *d*-spacings are measured from randomly-orientated nanocrystallites and can be indexed to the (C, E) (102) and (D) (004) planes of vaterite.

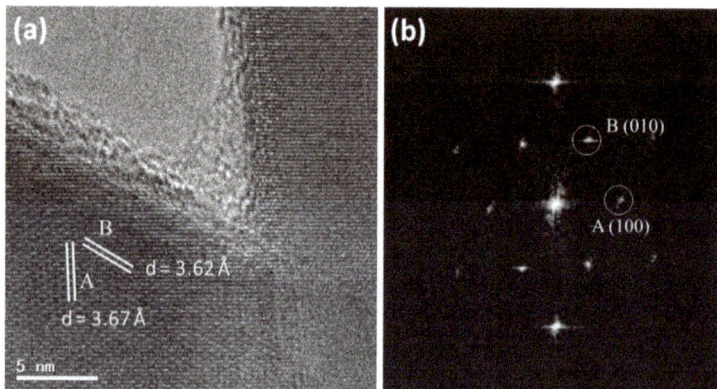

Figure 9. (**a**) HRTEM image of two cross-linked nanoneedles. (**b**) The corresponding FFT pattern, which is indexed to the hexagonal vaterite structure.

3. Discussion

Although a large number of SEM and HRTEM images of the spherulites, growing in limpet shells, have been obtained and analysed, the growth stages of these spherulites cannot be recognized undoubtedly. Drawing an evolution of the spherulites step-by-step as a function of time is difficult. However, a complete series of specimens during spherulite growth via biomimetic synthesis can be easily achieved. These results may help us to understand the time dependant formation of spherulites in a natural environment if the conditions of the latter can be simulated reasonably well.

Vaterite-type $CaCO_3$ spherulites have been biomimetically synthesized using gelatine as a biological substance [37]. The selection of gelatine is because it has an isoelectric point in a range of 4.7 to 5.2 [46], and has similar functions as a base of $CaCO_3$ deposition with the extracellular polymeric substance (EPS) matrix in biofilms commonly found in the environments for biomineralization. In an aqueous synthetic system containing gelatine (type B), $Ca(NO_3)_2 \cdot 4H_2O$, and urea, the growth of $CaCO_3$ and morphology evolution underwent several distinguished steps: Step 1, the formation of vaterite nanocrystallites, ~5 nm in diameter, embedded in a disordered gelatine matrix. Step 2, aggregation of the nanocrystallites into large spherical particles. Step 3, dipole field driven self-orientation of the nanocrystallites and the development of nanoneedles on a spherical core in a radial arrangement, forming spherulites. Step 4, development of a double-layer disk, i.e., two $CaCO_3$ plates sandwiching a gelatine layer. This disk supplies a relatively strong and mirror symmetric dipole field, guiding the orientations of all the nanocrystallites, leading to formations of an equatorial notch and "twin-cauliflower"-shaped particles [37]. It is obvious that, with the increase of growth time, the particle size and ordering of the nanocrystallites increase.

Many structural features observed from the spherulites in the limpet shells are similar with those in the synthetic spherulites [37], particularly the formation of the double-layer disks. We, therefore, assume that the principles of the formation of these spherulites in both biomineralization and biomimetic synthesis are similar. Dipole field interaction between the crystals is believed to be the most important factor governing the formation of the spherulites.

It has been known that hexagonal vaterite has a permanent dipole moment along the c-axis. The (001) surface is terminated with a Ca^{2+} layer and is positively charged, while the (00$\bar{1}$) surface is negatively charged with the top surface layer only containing CO_3^{2-} anions. If the distance between two isoelectric terminals is d, the dipole moment p can be calculated in Debye units:

$$p = \sum_i q_i d_i$$

where q_i is the charging magnitude and d_i is the distance of which the ith anion/cation held. Forces in a dipole field have correlation with distances and the charging level. For a dipole field, if a small particle is located at r_+ and r_- from positive and negative charges of the dipole, the electric potential $V(r)$ is:

$$V(r) = \frac{q}{4\pi\varepsilon_0 r_+} - \frac{q}{4\pi\varepsilon_0 r_-}$$

where π is a constant, and ε_0 is the vacuum permittivity.

If this particle is far enough from the dipole, i.e., $r^2 \approx r_+ r_-$ (r is the distance from the particle to the midpoint of a dipole. θ is the angle between the dipole and the direction from the midpoint of the dipole to the particle), the potential can be written as:

$$V(r) = \frac{1}{4\pi\varepsilon_0} \cdot \frac{p\cos\theta}{r^2}$$

In the dipole field, we have:

$$E(r) = \frac{1}{2\pi\varepsilon_0} \cdot \frac{p}{r^3}$$

where E(r) is the intensity of the dipole field at position r [47]. It can be seen from this equation that the field strength increases remarkably when r decreases, especially to the nanometer scale.

Based on the above discussion, we can now propose a multi-step formation mechanism for the naturally-occurring spherulites as presented in Figure 10.

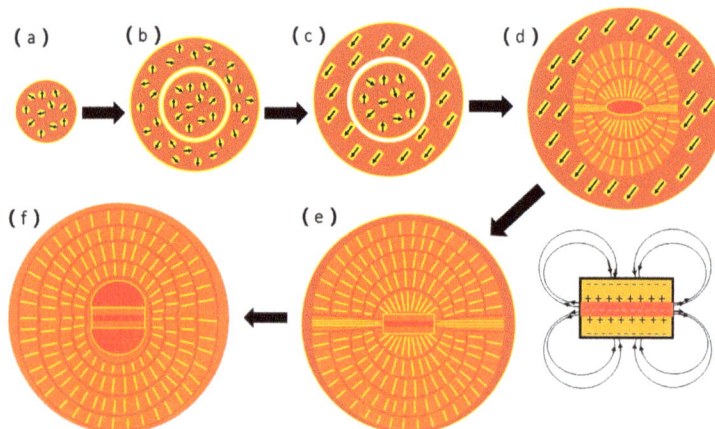

Figure 10. Schematic drawing of a proposed formation mechanism of spherulites grown in limpet shells. (**a**) Small spherical particles of inorganic/organic composite, in which vaterite nanocrystallites (yellow) are developed. (**b**) Separation of the core and the surroundings. (**c**) Formation of parallel vaterite short nanoneedles. (**d**) Formation of the central double-layer disk. (**e**) Formation of spherulites with two hemispheres separated by a notch. (Right bottom) A schematic drawing of the dipole field from the central disk. (**f**) Final stage of the formation of the spherulites.

Step 1: The biofilms with extracellular polymeric substance (EPS) in a gap of the M layer in the limpet shell are normally negatively charged and have a power to attract Ca^{2+} cations first, followed by attracting CO_3^{2-} anions, forming an inorganic/organic composite. When the concentration of the inorganic component is high, some spherical particles (50 to 200 nm in diameter) form, separating from the biofilms. Vaterite nanocrystallites develop from amorphous calcium carbonate (ACC) inside these spheres (Figure 10a). A typical experimental image of these spherical particles is shown in Figure 8.

Step 2: When the spheres increase their size and vaterite nanocrystallites grow further, an increase of the density of the central area is faster than the surroundings, and a core particle is separated from the surroundings (Figure 10b), as seen in Figure 7a.

Step 3: The vaterite nanocrystallites grow into short nanoneedles. A dipole field is created from each nanoneedle when it forms a regular shape. An interaction between these nanoneedles force them to self-orientate into a parallel arrangement inside the spherical particles (Figure 10c). The corresponding SEM image is shown in Figure 7d.

Step 4: The core spherical particle undergoes surface re-crystallization into a double-layer disk sandwiching a biological sheet. The formation of such a double-layer disk via an inorganic/organic phase separation was observed in the biomimetic synthesis of $CaCO_3$ in the presence of gelatin [37], and in mirror symmetric growth of ZnO nanoneedles/plates [48,49]. It is also regarded as a good example of two-dimensional reversed crystal growth, a non-classical crystal growth route demonstrating a crystal growth direction from a particle surface to its core [50–52]. More importantly, the central biological sheet is normally negatively charged. Therefore, the inner surfaces of both vaterite layers are positively charged and the outer surfaces must be negatively charged. Such a structure offers a strong dipole field, which will guide all the nanocrystallites to adjust their orientations along the field force lines (Figure 10d). The corresponding experimental observation is shown in Figure 2d.

Step 5: When all the nanocrystallites line up into nanoneedles and, later, mesorods, and all these 1D particles lie along the dipole field force lines created from the central double-layer disk, a spherulite forms with two hemispheres separated by a notch where the field strength is the weakest (Figure 10e), as seen in SEM images of Figure 2a,c. Both the spaces between the mesorods and between the hemispherical layers are filled by biological substances. During the above-mentioned process, a phase transformation from vaterite to aragonite takes place. However, the calcite phase was not detected from spherulites, which is the main crystalline phase in other parts of the limpet shells. Another interesting phenomenon is that, although the particles are divided into two hemispheres connected only by a central core, the number of the mesorod layers developed in both hemispheres are the same and the thicknesses of any face-to-face corresponding mesorod layers are uniform (Figure 5a). The growth is, therefore, mirror symmetric.

Step 6: Finally, as depicted in Figure 10f, the density of the spherulites increases. The corresponding mesorod layers in the two hemispheres join together across the notch, which is not visible on the outer surface (Figure 5b). All the mesorod layers in the hemispheres can match the opposite ones and perfectly connect each other. On the other hand, on the both sides of the core disk, further phase separation leads to two hemispherical spaces filled by biological substances, as shown by two dark areas in the core in Figure 5a,b, as well as Figure 6.

Nevertheless, the mechanism presented in Figure 10 is still an assumption, rather than being established unambiguously. Future research is needed to prove all the steps in this mechanism.

4. Experimental

4.1. Specimen Collection and Preparation

Limpets were collected at East Sands, St. Andrews, Scotland. They were sacrificed and the shells were obtained. The limpet shells were embedded in epoxy resin and were easily cut with a low-speed wheel diamond saw. The samples were then polished with sandpaper (1200 grit). A weak acidic solution (5 mmol/L EDTA, pH = 4.0, etching time 30 min) was used for the pre-treatment of the sample, in order to remove the contaminants and fragments on the surface of the sample. The samples were washed with distilled water and acetone. Ultrasonic cleaning was also performed for 30 min.

4.2. Characterization

The SEM images of the polished specimen surfaces were taken on a JSM-5600 scanning electron microscope and a JSM-6700F field-emission gun microscope with an accelerating voltage of 5 kV. Samples were coated with a thin gold film by using a sputter coater to improve the conducting property to avoid electron beam charging. Energy dispersive X-ray (EDX) spectroscopy elemental mapping was performed on an Oxford INCA system (Oxford Instruments, Abingdon, UK) equipped on the JSM-5600.

Powder X-ray diffraction (XRD) patterns were collected on a PANalytical Empyrean diffractometer (PANalytical, Levi, Finland) using copper Kα radiation with λ = 1.5418 Å. The results of PXRD were processed by using Highscore Plus software (PANalytical, Levi, Finland).

For TEM and HRTEM investigations, samples were prepared in either powder or thin plates. The latter was made by using a focused ion beam facility on a FEI Scios dual-beam microscope. TEM and HRTEM images were taken on a JEOL JEM-2011 electron microscope (Jeol, Tokyo, Japan), operated at 200 kV with a Gatan 794 CCD camera.

5. Conclusions

In the present work, the hierarchical microstructures of spherulites grown in limpet shells have been investigated by using SEM and HRTEM. In particular, spherulites at different growth stages were examined. From the results, a formation mechanism of the spherulites has been proposed. It has been assumed that a mirror symmetric dipole field associated with the double-layer disks in the cores is the driving force guiding the self-orientation of the nanocrystallites and, therefore, the morphology evolution. To achieve this, the nanocrystallites embedded in a biological substance matrix must be able to rotate and shift locally. This work may shed light on the research of biomineralization in many other biological systems and biomimetic synthesis of many other crystals.

Supplementary Materials: The following are available online at www.mdpi.com/2073-4352/7/10/319/s1, Figure S1: SAED pattern recorded from a large area in a mesorod, covering polycrystalline area containing self-orientated nanocrystallites and single crystalline region. The corresponding HRTEM images are shown in Figure 4b,c in the paper.

Acknowledgments: The authors would like to thank EPSRC for financial support for the FEG-SEM equipment (EP/F019580/1) and FEI Scios dual-beam microscope (EP/L017008/01).

Author Contributions: Shitao Wu and Chang-Yang Chiang performed most experiments. Wuzong Zhou designed the project and suggested experiments. All the authors contributed to the analysis of the results. Shitao Wu and Wuzong Zhou wrote the paper.

Conflicts of Interest: The authors declare no conflict of interest.

References

1. Lin, A.Y.; Meyers, M.A. Growth and structure in abalone shell. *Mater. Sci. Eng. A-Struct.* **2005**, *390*, 27–41. [CrossRef]
2. Lin, A.Y.; Chen, P.Y.; Meyers, M.A. The growth of nacre in the abalone shell. *Acta Biomater.* **2008**, *4*, 131–138. [CrossRef] [PubMed]
3. Sun, J.; Bhushan, B. Hierarchical structure and mechanical properties of nacre: A review. *RSC Adv.* **2012**, *2*, 7617–7632. [CrossRef]
4. Gibbs, S.J.; Poulton, A.J.; Bown, P.R.; Daniels, C.J.; Hopkins, J.; Young, J.R.; Jones, H.L.; Thiemann, G.J.; O'Dea, S.A.; Newsam, C. Species-specific growth response of coccolithophores to palaeocene-eocene environmental change. *Nat. Geosci.* **2013**, *6*, 218–222. [CrossRef]
5. Politi, L.; Talmon, A.; Klein, E.; Weiner, S.; Addadi, L. Sea urchin spine calcite forms via a transient amorphous calcium carbonate phase. *Science* **2004**, *306*, 1161–1164. [CrossRef] [PubMed]
6. Kawano, J.; Shimobayashi, N.; Miyake, A.; Kitamura, M. Precipitation diagram of calcium carbonate polymorphs: Its construction and significance. *J. Phys. Condens. Matter* **2009**, *21*, 425102. [CrossRef] [PubMed]
7. Spanos, N.; Koutsoukos, P.G. The transformation of vaterite to calcite: Effect of the conditions of the solutions in contact with the mineral phase. *J. Cryst. Growth* **1998**, *191*, 783–790. [CrossRef]
8. Berner, R.A. The role of magnesium in the crystal growth of calcite and aragonite from sea water. *Geochim. Cosmochim. Acta* **1975**, *39*, 489–504. [CrossRef]
9. Nassif, N.; Gehrke, N.; Pinna, N.; Shirshova, N.; Tauer, K.; Antonietti, M.; Colfen, H. Synthesis of stable aragonite superstructures by a biomimetic crystallization pathway. *Angew. Chem. Int. Ed. Engl.* **2005**, *44*, 6004–6009. [CrossRef] [PubMed]
10. Rousseau, M.; Lopez, E.; Stempfle, P.; Brendle, M.; Franke, L.; Guette, A.; Naslain, R.; Bourrat, X. Multiscale structure of sheet nacre. *Biomaterials* **2005**, *26*, 6254–6262. [CrossRef] [PubMed]
11. Addadi, L.; Joester, D.; Nudelman, F.; Weiner, S. Mollusk shell formation: A source of new concepts for understanding biomineralization processes. *Chem.-Eur. J.* **2006**, *12*, 980–987. [CrossRef] [PubMed]
12. Nudelman, F.; Gotliv, B.A.; Addadi, L.; Weiner, S. Mollusk shell formation: Mapping the distribution of organic matrix components underlying a single aragonitic tablet in nacre. *J. Struct. Biol.* **2006**, *153*, 176–187. [CrossRef] [PubMed]
13. Niederberger, M.; Colfen, H. Oriented attachment and mesocrystals: Non-classical crystallization mechanisms based on nanoparticle assembly. *Phys. Chem. Chem. Phys.* **2006**, *8*, 3271–3287. [CrossRef] [PubMed]
14. Suzuki, M.; Kameda, J.; Sasaki, T.; Saruwatari, K.; Nagasawa, H.; Kogure, T. Characterization of the multilayered shell of a limpet, lottia kogamogai (mollusca: Patellogastropoda), using SEM-EBSD and FIB-TEM techniques. *J. Struct. Biol.* **2010**, *171*, 223–230. [CrossRef] [PubMed]

15. Soldati, A.L.; Jacob, D.E.; Wehrmeister, U.; Hofmeister, W. Structural characterization and chemical composition of aragonite and vaterite in freshwater cultured pearls. *Mineral. Mag.* **2008**, *72*, 579–592. [CrossRef]
16. Wehrmeister, U.; Jacob, D.E.; Soldati, A.L.; Häger, T.; Hofmeister, W. Vaterite in freshwater cultured pearls from China and Japan. *J. Gemm.* **2007**, *31*, 269–276. [CrossRef]
17. Mann, S. *Biomineralization: Principles and Concepts in Bioinorganic Materials Chemistry*; Oxford University Press: New York, NY, USA, 2001.
18. Wray, J.; Daniels, F. Precipitation of calcite and aragonite. *J. Am. Chem. Soc.* **1957**, *79*, 2031–2034. [CrossRef]
19. Ogino, T.; Suzuki, T.; Sawada, K. The formation and transformation mechanism of calcium carbonate in water. *Geochim. Cosmochim. Acta* **1987**, *51*, 2757–2767. [CrossRef]
20. Di Lorenzo, M.L. Spherulite growth rates in binary polymer blends. *Prog. Polym. Sci.* **2003**, *28*, 663–689. [CrossRef]
21. Miao, B.; Wood, D.O.N.; Bian, W.; Fang, K.; Fan, M.H. Structure and growth of platelets in graphite spherulites in cast iron. *J. Mater. Sci.* **1994**, *29*, 255–261. [CrossRef]
22. Magill, J.H.; Plazek, D.J. Physical properties of aromatic hydrocarbons. II. Solidification behavior of 1,3,5-tri-α-naphthylbenzene. *J. Chem. Phys.* **1967**, *46*, 3757–3769. [CrossRef]
23. Morse, H.W.; Donnay, J.D.H. Optics and structure of three-dimensional spherulites. *Am. Mineral.* **1936**, *21*, 391–426.
24. Gardner, J.E.; Befus, K.S.; Watkins, J.; Hesse, M.; Miller, N. Compositional gradients surrounding spherulites in obsidian and their relationship to spherulite growth and lava cooling. *Bull. Volcanol.* **2012**, *74*, 1865–1879. [CrossRef]
25. Wright, V.P. Lacustrine carbonates in rift settings: The interaction of volcanic and microbial processes on carbonate deposition. *J. Geol. Soc. Lond. Spec. Pub.* **2012**, *370*, 39–47. [CrossRef]
26. Zelenkova, M.; Sohnel, O.; Grases, F. Ultrafine structure of the hydroxyapatite amorphous phase in noninfectious phosphate renal calculi. *Urology* **2012**, *79*, 961–966. [CrossRef] [PubMed]
27. Granasy, L.; Pusztai, T.; Tegze, G.; Warren, J.A.; Douglas, J.F. Growth and form of spherulites. *Phys. Rev. E* **2005**, *72*, 011605. [CrossRef] [PubMed]
28. Banfield, J.F.; Welch, S.A.; Zhang, H.; Ebert, T.T.; Penn, R.L. Aggregation-based crystal growth and microstructure development in natural iron oxyhydroxide biomineralization products. *Science* **2000**, *289*, 751–754. [CrossRef] [PubMed]
29. Tracy, S.L.; François, C.J.P.; Jennings, H.M. The growth of calcite spherulites from solution: I. Experimental design techniques. *J. Cryst. Growth* **1998**, *193*, 374–381. [CrossRef]
30. Tracy, S.L.; Williams, D.A.; Jennings, H.M. The growth of calcite spherulites from solution II. Kinetics of formation. *J. Cryst. Growth* **1998**, *193*, 382–388. [CrossRef]
31. Rodriguez-Blanco, J.D.; Sand, K.K.; Benning, L.G. ACC and vaterite as intermediates in the solution-based crystallization of CaCO₃. In *New Perspectives on Mineral Nucleation and Growth*; Van Driessche, A.E.S., Kellermeier, M., Benning, L.G., Gebauer, D., Eds.; Springer: Cham, Switzerland, 2017; pp. 93–111.
32. Addadi, L.; Raz, S.; Weiner, S. Taking advantage of disorder: Amorphous calcium carbonate and its roles in biomineralization. *Adv. Mater.* **2003**, *15*, 959–970. [CrossRef]
33. Tobler, D.J.; Rodriguez-Blanco, J.D.; Sørensen, H.O.; Stipp, S.L.S.; Dideriksen, K. Effect of pH on amorphous calcium carbonate structure and transformation. *Cryst. Growth Des.* **2016**, *16*, 4500–4508. [CrossRef]
34. Bots, P.; Rodriguez-Blanco, J.D.; Benning, L.G.; Shaw, S. Mechanistic insights into the crystallization of amorphous calcium carbonate to vaterite. *Cryst. Growth Des.* **2012**, *12*, 3806–3814. [CrossRef]
35. Mercedes-Martín, R.; Rogerson, M.R.; Brasier, A.T.; Vonhof, H.B.; Prior, T.J.; Fellows, S.M.; Reijmer, J.J.G.; Billing, I.; Pedley, H.M. Growing spherulitic calcite grains in saline, hyperalkaline lakes: Experimental evaluation of the effects of Mg-clays and organic acids. *Sediment. Geol.* **2016**, *335*, 93–102. [CrossRef]
36. Rodriguez-Navarro, C.; Jimenez-Lopez, C.; Rodriguez-Navarro, A.; Gonzalez-Muñoz, M.T.; Rodriguez-Gallego, M. Bacterially mediated mineralization of vaterite. *Geochim. Cosmochim. Acta* **2007**, *71*, 1197–1213. [CrossRef]
37. Greer, H.F.; Liu, M.H.; Mou, C.Y.; Zhou, W.Z. Dipole field driven morphology evolution in biomimetic vaterite. *CrystEngComm* **2016**, *18*, 1585–1599. [CrossRef]
38. Andreassen, J.P.; Flaten, E.M.; Beck, R.; Lewis, A.E. Investigations of spherulitic growth in industrial crystallization. *Chem. Eng. Res. Des.* **2010**, *88*, 1163–1168. [CrossRef]

39. Andreassen, J.P. Formation mechanism and morphology in precipitation of vaterite—nano-aggregation or crystal growth? *J. Cryst. Growth* **2005**, *274*, 256–264. [CrossRef]

40. Zhong, C.; Chu, C.C. On the origin of amorphous cores in biomimetic CaCO$_3$ spherulites: New insights into spherulitic crystallization. *Cryst. Growth Des.* **2010**, *10*, 5043–5049. [CrossRef]

41. Zhou, G.T.; Yao, Q.Z.; Ni, J.; Jin, G. Formation of aragonite mesocrystals and implication for biomineralization. *Am. Mineral.* **2009**, *94*, 293–302. [CrossRef]

42. Nan, Z.; Shi, Z.; Yan, B.; Guo, R.; Hou, W. A novel morphology of aragonite and an abnormal polymorph transformation from calcite to aragonite with PAM and CTAB as additives. *J. Colloid. Interface. Sci.* **2008**, *317*, 77–82. [CrossRef] [PubMed]

43. Kamhi, S.R. On the structure of vaterite CaCO$_3$. *Acta Crystallogr.* **1963**, *16*, 770–772. [CrossRef]

44. Greer, H.F.; Zhou, W.Z. Electron diffraction and HRTEM imaging of beam sensitive materials. *Crystallogr. Rev.* **2011**, *17*, 163–185. [CrossRef]

45. Gago-Duport, L.; Briones, M.J.; Rodriguez, J.B.; Covelo, B. Amorphous calcium carbonate biomineralization in the earthworm's calciferous gland: Pathways to the formation of crystalline phases. *J. Struct. Biol.* **2008**, *162*, 422–435. [CrossRef] [PubMed]

46. Bauermann, L.P.; Del Campo, A.; Bill, J.; Aldinger, F. Heterogeneous nucleation of ZnO using gelatin as the organic matrix. *Chem. Mater.* **2006**, *18*, 2016–2020. [CrossRef]

47. Walker, J.D. *Fundamentals of Physics Extended*; Wiley: New York, NY, USA, 2010.

48. Liu, M.-H.; Tseng, Y.-H.; Greer, H.F.; Zhou, W.Z.; Mou, C.Y. Dipole field guided orientated attachment of nanocrystals to twin-brush ZnO mesocrystals. *Chem. Eur. J.* **2012**, *18*, 16104–16113. [CrossRef] [PubMed]

49. Greer, H.F.; Zhou, W.Z.; Liu, M.-H.; Tseng, Y.-H.; Mou, C.-Y. The origin of ZnO twin crystals in bio-inspired synthesis. *CrystEngComm* **2012**, *14*, 1247–1255. [CrossRef]

50. Chen, X.Y.; Qiao, M.H.; Xie, S.H.; Fan, K.N.; Zhou, W.Z.; He, H.Y. Self-construction of core-shell and hollow zeolite analcime icositetrahedra: A reversed crystal growth process via oriented aggregation of nanocrystallites and recrystallization from surface to core. *J. Am. Chem. Soc.* **2007**, *129*, 13305–13312. [CrossRef] [PubMed]

51. Zhou, W.Z. Reversed crystal growth: Implications for crystal engineering. *Adv. Mater.* **2010**, *22*, 3086–3092. [CrossRef] [PubMed]

52. Yao, J.F.; Li, D.; Zhang, X.Y.; Kong, C.H.; Yue, W.B.; Zhou, W.Z.; Wang, H.T. Cubes of zeolite A with an amorphous core. *Angew. Chem. Int. Ed.* **2008**, *47*, 8397–8399. [CrossRef] [PubMed]

crystals

MDPI

Article

Modulating Nucleation by Kosmotropes and Chaotropes: Testing the Waters

Ashit Rao [1,2,3,*], Denis Gebauer [1,*] and Helmut Cölfen [1]

[1]   Department of Chemistry, Universitätsstr. 10, University of Konstanz, Konstanz 78464, Germany;
      helmut.coelfen@uni-konstanz.de
[2]   Freiburg Institute for Advanced Studies, Albert-Ludwigs-Universität Freiburg, Freiburg 79104, Germany
[3]   Centre for Biosystem Analysis, Albert-Ludwigs-Universität Freiburg, Freiburg 79104, Germany
*   Correspondence: Ashit.Rao@frias.uni-freiburg.de (A.R.); Denis.Gebauer@uni-konstanz.de (D.G.)

Received: 24 August 2017; Accepted: 4 October 2017; Published: 6 October 2017

Abstract: Water is a fundamental solvent sustaining life, key to the conformations and equilibria associated with solute species. Emerging studies on nucleation and crystallization phenomena reveal that the dynamics of hydration associated with mineral precursors are critical in determining material formation and growth. With certain small molecules affecting the hydration and conformational stability of co-solutes, this study systematically explores the effects of these chaotropes and kosmotropes as well as certain sugar enantiomers on the early stages of calcium carbonate formation. These small molecules appear to modulate mineral nucleation in a class-dependent manner. The observed effects are finite in comparison to the established, strong interactions between charged polymers and intermediate mineral forms. Thus, perturbations to hydration dynamics of ion clusters by co-solute species can affect nucleation phenomena in a discernable manner.

Keywords: crystallization; nucleation; kosmotropes; chaotropes; small molecules; hydration

1. Introduction

Water is one of the most abundant molecules in the universe [1]. It affects global processes such as erosion and climate as well as phenomena at atomic length scales such as molecular configurations and interactions. As a ubiquitous cellular constituent, water molecules influence biomolecular conformations and interactions [2,3]. However, the influences of hydration and bulk water on chemical processes are not completely understood. Historically, certain salts have been linked with an ability to influence water structure, and thereby the conformational states of co-solute macromolecules. This gradually led to the nomenclature of solute additives as either structure-breakers or structure-makers, corresponding to their respective stabilizing or disruptive effect on the short-range order i.e., hydrogen bonding networks in the liquid [2,4]. However, convincing evidence for a significant perturbation of bulk water structure by solutes still remains lacking.

Alternate explanations are provided for the observed effects of ionic species on macromolecules [3]. For instance, direct interactions of ions with macromolecules and their first shell of hydration play key roles in the hydrophobic collapse and solubility of poly(N-isopropylacrylamide) [5]. Simulation studies also show that ions alter hydrogen bonding, salt bridges, and hydrophobic interactions that underlie macromolecular conformations [6,7]. These studies suggest that the Hofmeister series emerges from interactions of the ions with macromolecules and the associated hydration, and not due to structural perturbations to the bulk solvent. Therefore the configurational and chemical properties of solute macromolecules are important factors that determine the consequences of ionic interactions [3,8,9].

Certain small organic molecules are also implicated in affecting the hydration of co-solutes. For example, organisms thriving in extreme habitats modify solvent effects by sequestering high solute contents [10]. Members of this solute family (compensatory kosmotropes or osmolytes [11])

are generally compatible with biochemical processes and provide a mechanism for the evasion of osmotic stress without extensive covalent modifications of biomacromolecules [10]. This is essential for the survival of organisms including bacterial and fungal spores, rotifers, and tardigrades [12,13]. The mechanistic functions of certain osmolytes (e.g., glycerol and betaine) may be due to their exclusion from the immediate vicinity of co-solutes [14,15]. These kosmotropes may also decrease the solubility of hydrophobic, potentially toxic molecules by enhancing their aggregation [16]. Being polar molecules with a negligible net charge, kosmotropes prefer forming hydrogen bonds with water molecules and therefore are excluded from the hydration shell of hydrophobic patches on macromolecules. On these lines, a 'solvophobic thermodynamic force' is described for the unfavorable interactions between the osmolytes and peptide backbone that increase the free energy for macromolecular denaturation [17]. Another hypothesis for the activity of kosmotropes is the occurrence of microdomains of high and low density states of water in which osmolytes (kosmotropes) and chaotropes preferentially partition. In this scenario, the thermodynamic cost for altering the equilibrium between the density states is related to the conformational stability of macromolecular co-solutes [18].

In comparison to kosmotropes, in aqueous solutions of chaotropes, the solubility of non-polar solutes is enhanced. For instance, urea and guanidine significantly increase the solubility of biomolecules; however at the cost of denaturation, i.e., a loss of native macromolecular structure [19,20]. These small molecules interact with co-solutes and weaken interactions with solvent molecules. This leads to destabilized native conformations and an increased water-accessible surface area of macromolecules [16,21]. A preferential accumulation of chaotropic metabolites is also shown to support the growth of certain microbes at low temperatures [22]. Thus the effects of chaotropes and kosmotropes on the conformation and hydration of co-solutes arises from their preferential partitioning to either the bulk solvent phase or the water-solute interfaces and therefore is determined by the physicochemical nature of hydrated co-solutes [23]. Given the fundamental roles of these interactions in biomolecular structure and function, their impacts on nucleation and crystallization behavior in nature also require elucidation.

Of the natural minerals, calcium carbonate ($CaCO_3$) is produced at a rate of 5 billion tons per year in the oceans and is a key industrial material on account of its role as a filler and scalant material [24–26]. It is a crucial raw material for the production of Portland cement (about 4 Gt manufactured per year). With advances in the resolution of analytical techniques, several key aspects of mineral nucleation and growth including pre-nucleation clusters (PNCs) and oriented attachment have been identified [27–32]. In these processes, the role of water as a bulk solvent and hydration molecule is fundamentally important. The formation of ion-clusters from free species requires a certain release of ion-associated hydration [28,29]. During the phase separation of hydrated PNCs, the dynamics of the water hydrogen network are affected and allow localizing a binodal limit for liquid-liquid demixing. This process yields dense calcium carbonate droplets, which are precursors to solid amorphous calcium carbonate (ACC) [33–35]. Recently, it was shown that different background ions influence PNC stabilities as well as ACC solubilities, which was rationalized by effects of the spectator ions on the hydration shells [36]. Considering the lower density of PNCs in comparison to that of liquid/gel-like and solid ACC, a significant ion cluster-associated hydration endures in the nucleated phase [37]. These intermediate mineral forms are characterized by distinct water contents. Stable forms of biogenic and synthetic ACC also contain structural water [38–41]. During the process of crystallization, the ACC undergoes steps of hydration loss and follows an energetically downhill sequence from a short-lived anhydrous form to crystalline phases [42,43]. This transformation process is significantly affected by the water content of the bulk solvent, affecting the polymorph selectivity of $CaCO_3$ [44]. In the absence of additives stabilizing crystalline surfaces, the solvent conditions also determine the reconstruction of high energy faces [45]. These studies reflect a dynamic equilibrium between free (bulk) and ion-associated (hydration/structural) water during mineral formation, the importance of which is reflected by the distinct nature of the H-bond network of hydration shells in terms of density, structure, and dynamics [46–48]. Thus the bulk and mineral-associated solvent molecules play critical roles from the primary to final stages of mineralization, encompassing PNCs, liquid, and amorphous intermediates and crystals.

The roles of solute-associated and bulk solvent as "additives" are also demonstrated for several material systems. Considering calcium sulphate, solvent polarity, i.e., its water-withdrawing ability, has a direct bearing on the polymorph selection of bassanite and gypsum [49]. For organic crystals, the solvent composition also determines the selective nucleation of polymorphic forms [50]. For oriented attachment processes, the solvent modulates interactions with organic ligands and also provides a certain degree of particle movement that enables co-orientation and coalescence [31,51–54]. Thus the solvent phase commonly affects nucleation and crystallization processes as well as the subsequent material properties including crystallography, morphology, and polymorph.

In view of certain co-solutes modulating molecular hydration and also the significant contributions of solvent parameters towards nucleation and crystal growth, the aim of this study is to understand, "Do chaotropes and kosmotropes affect mineral nucleation?" For this purpose, small molecules typically classified as chaotropes and kosmotropes are quantitatively evaluated as additives during the nucleation of $CaCO_3$. The additives tested are betaine, ectoine, trehalose, sorbitol, mannitol, glycine, urea, thiourea, and guanidium. With reports of the stereoisomers of mono- and oligo-saccharides in chaotropic and kosmotropic behavior [55], D- and L-forms of sugars are also investigated. The analytical approach involves potentiometric titrations performed by dosing calcium chloride into a solution of an additive dissolved in carbonate buffer. During the course of the experiment, the pH is kept constant by the counter-titration of a base. Similar experiments have been previously performed to assess the effects of diverse additives including synthetic polymers [26], biomolecules such as amino acids [56], carbohydrates, polysaccharides [57,58], proteins [59], and peptides [60] as well as inorganic species such as silica and magnesium [61–64]. By employing ion-selective electrodes and precise reaction conditions, the diverse roles of additives on different aspects of mineral nucleation including the stability of PNCs, inhibition or promotion of nucleation, and the nature of the nucleated phase are demonstrated.

2. Results

2.1. On Kosmotropes and Chaotropes

Dataset analyses for the time developments of free Ca^{2+} ion and pH are performed as described in previous literature [28,56,57]. For titrations performed in carbonate buffer, the detected free Ca^{2+} ion concentration increases linearly with time at a rate significantly lower than that of Ca^{2+} ion addition to water. This is due to the association of Ca^{2+} ions and carbonate species that result in diminished detected free Ca^{2+} ion contents [28]. The linear slope corresponding to the pre-nucleation regime describes the formation and stability of PNCs. A lower slope indicates an equilibrium shift towards cluster-associated Ca^{2+} ions i.e., bound Ca^{2+} and vice versa. For example, the evolution of free Ca^{2+} ions as well as the added base contents to maintain constant pH in absence or presence of different additives are represented in Figure 1. The constant pH level is required to quantitatively compare the effects of the chao-/kosmotropic additives during the early stages of calcium carbonate formation because it ensures a constant carbonate/bicarbonate ratio and removes corresponding, otherwise disturbing effects on supersaturation and ionic speciation.

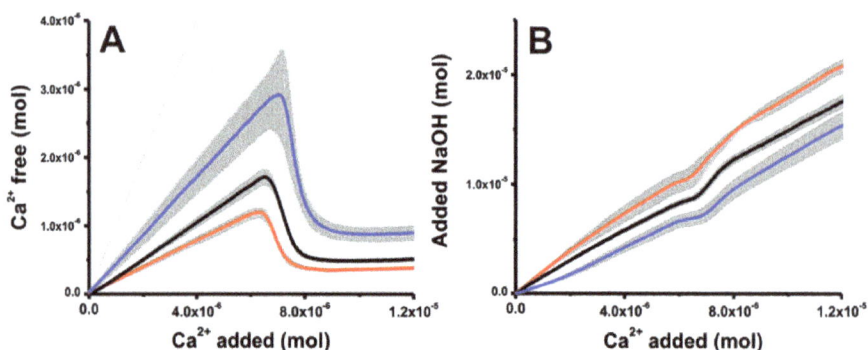

Figure 1. Developments of (**A**) free Ca^{2+} ions and (**B**) added NaOH contents for titrations performed at pH 9.75 in absence of additives (black) and the presence of either 10 mM urea (red) or trehalose (blue) in relation to the total Ca^{2+} ion contents. Zones shaded grey represent ±1 standard deviation for triplicate experiments. (**A**) The dotted line represents the development of total amount of Ca^{2+} ions dosed into the buffer.

As examples of kosmotropes and chaotropes in the field of cellular stress biology [65], the effects of trehalose and urea are compared to reference experiments at pH 9.75. These additives do not influence the time required for mineral nucleation but do significantly affect the equilibrium between free and carbonate-associated Ca^{2+} ions, as evident from the slope in the prenucleation regime before the sudden Ca^{2+} concentration drop upon nucleation. From Figure 1A, it is evident that trehalose leads to free Ca^{2+} ion contents higher than those in the reference experiments. On the other hand, urea stabilizes ion-clusters, presenting significant lower amounts of free Ca^{2+} ions. These observations are also reflected by the amount of base required for maintaining constant pH conditions (Figure 1B). This amount indicates the carbonate ions bound in the different stages of the experiment [28]. For instance, urea, as an effective stabilizer of ion-clusters, necessitates larger quantities of base solutions to maintain constant pH conditions. The stronger and weaker binding in presence of urea and trehalose, respectively, is seen in two independent measurements (calcium ion selective electrode and pH titration), and thus, the effect of the additives cannot be an effect of specific electrode artifacts. Therefore, given its sensitivity and quantitative nature, this methodology is suitable for addressing the effects of small organic molecules on mineral nucleation.

Figure 2A represents the slopes of the pre-nucleation regime in the absence and presence of additives at pH 9.0 and 9.75. Trehalose, sorbitol, ectoine, mannitol, and betaine induce increases in the pre-nucleation slope in comparison to the relative reference experiments at both pH values. At pH 9.0, this corresponds to percentage increases of about 87, 44, 29, 26, and 20%, respectively. At higher pH (9.75), these effects are more pronounced, leading to corresponding increases of about 41, 56, 66, 109, and 39% in the pre-nucleation slope. In this regard, trehalose and mannitol suppress ion-association and are effective pH-dependent destabilizers of PNCs. In view of the pre-nucleation regime, opposite outcomes are elicited by chaotropes. This is evident from the decreased slope of the pre-nucleation regimes corresponding to 30, 22, and 14% in presence of urea, guanidium and thiourea, respectively at pH 9.75. These effects are diminished at pH 9.0, however, urea retains its effect as a PNC stabilizer at pH 9.0 and 9.75. In view of the observed equilibrium shifts between free and bound ion species, the distinct bicarbonate and carbonate ratio appear to impart a pH-dependent physicochemistry to the PNCs, thus consequently determining additive-ion and –ion cluster interactions.

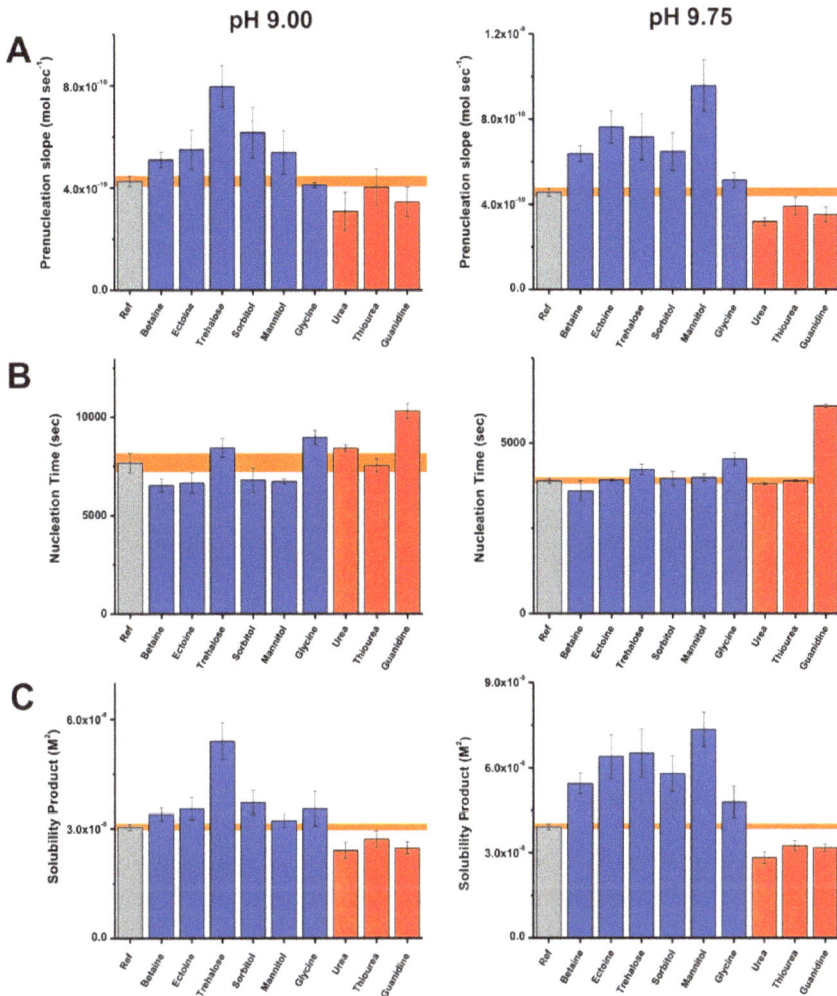

Figure 2. Bar plots representing the early stages of $CaCO_3$ nucleation in terms of the (**A**) pre-nucleation slope, (**B**) time required for nucleation and (**C**) solubility of initially nucleated phase at pH 9.0 (left) and 9.75 (right) in the presence of kosmotropic (blue) and chaotropic (red) additives and also in reference (Ref, gray) experiments. Error bars depict ± one standard deviation and corresponding values for reference experiments are shaded (orange).

The effects of the small molecules on the time required for mineral nucleation are represented in Figure 2B. As introduced in previous studies [56,57], we apply a scaling factor (F) defined as the quotient of the average nucleation time in presence of an additive and that of the corresponding reference. At pH 9.75, several additives exhibit F values close to unity viz. no observed effect on nucleation time. As exceptions, glycine and guanidine inhibit nucleation, with corresponding F values of 1.03 and 1.1 at pH 9.0 and 1.2 and 1.6 at pH 9.75, all respectively. Previously, the inhibition of nucleation was attributed to the colloidal stabilization of nanoscopic precursors against aggregation [61], but it might also be related to an influence of the additives on the size of pre-nucleation clusters [66]. In either case, the increasing extent of inhibition at higher pH might suggest the involvement of electrostatic interactions driven by pH-related additive speciation. In the case of glycine, this is supported by the predominance of its base form above pH 9.5 (Figure S1). The inhibitory effects introduced by these molecules are small, in comparison to those in the presence of polymeric additives such as poly(acrylic acid), poly(aspartic acid), carboxymethyl cellulose, and heparin [25,26,56,57]. Certain small molecules promote mineral nucleation at pH 9.0. The corresponding F values presented by kosmotropes such as betaine, ectoine, and mannitol are 0.76, 0.77, and 0.78. It is intriguing that although minor, a nucleation promoting effect is induced by these small molecules. A similar effect is noted in the presence of D-arabinose, D-galactose, and sucrose [57]. In the scope of this study, the exact reasons underlying the weak nucleation promoting effect cannot be identified. However, we speculate that the promotion of mineral nucleation is related to the liquid–liquid phase separation, during which PNCs become nanodroplets that can subsequently yield either ACC or crystalline particles.

Potentiometric titrations also elucidate the nature of mineral products [26,28]. The estimated solubility products of the nucleated CaCO$_3$ phase are depicted in Figure 2C. Overall, the kosmotropes and chaotropes lead to the formation of more and less stable phases after nucleation in relation to reference experiments, respectively. The reference experiments present solubility product values of 3.0 × 10^{-8} and 3.8 × 10^{-8} M^2 after nucleation at pH 9.0 and 9.75, in agreement with previous reports within experimental accuracy [28,56,57]. As a general trend, mineral products nucleated in the presence of kosmotropes viz. betaine, ectoine, trehalose, sorbitol and mannitol exhibit solubility products higher than those of the reference experiments. The most soluble mineral products are formed in presence of trehalose and mannitol corresponding to mean values of 5.4 × 10^{-8} and 7.4 × 10^{-8} M^2 at pH 9.0 and 9.75, respectively. On the other hand, urea, thiourea, and guanidine induce less soluble mineral products. At pH 9.0, the corresponding average values of the solubility product are between 2.4 × 10^{-8} and 2.7 × 10^{-8} M^2 indicating the presence of a crystalline polymorph (possibly vaterite). At pH 9.75, these values range between 2.8 × 10^{-8} and 3.2 × 10^{-8} M^2, which are lower than the solubility product of mineral particles nucleated in reference experiments (3.8 × 10^{-8} M^2). This reflects the presence of either a relatively short-lived amorphous product or a phase with lower solubility (such as proto-calcitic ACC or vaterite), which are distinct from the proto-vaterite ACC phase produced in the reference experiment at pH 9.75 [38]. Thus, kosmotropes and chaotropes have distinct influences on the solubility product of the mineral phase nucleated.

In order to explore possible relations between the observed effects of additives on mineral nucleation and their activity as promotors or disruptors of macromolecular conformation, a scatter plot for the pre-nucleation slopes, which indicate the amount of formed PNCs (the lower the slope the more PNCs are formed) and post-nucleation solubility products is presented (Figure 3). Additives that induce equilibrium shifts towards ion-association, also stimulate the formation of nucleated phase with lower solubility values. This relation has been previously identified with respect to conditions of pH and also in the presence of certain additives [28,67,68]. The applied color scheme represents a chao/kosmotropicity scale based on a previous systematic study wherein (i) chaotropes are defined as solutes that induce conformational disorder in macromolecules by either weakening water-macromolecule interactions or associating non-covalently with macromolecules and (ii) kosmotropes are solutes that preferentially hydrogen bond with water molecules and promote intermolecular interactions [69]. Within the scope of this definition, the effects of the chaotropes on mineral nucleation might be due to direct interactions with ions and ion-cluster species. On the other hand, the kosmotropes suppress ion-association and promote nucleation products with higher solubilities on account of the changes in the hydration state of the inorganic species. In validation of the findings that additive-controlled equilibrium shifts towards ion-association also stimulate the nucleation of mineral phases with lower solubility values (Figure 3), gas-diffusion mineralization experiments are performed (Figure S2). Ectoine and betaine lead to the formation of a mixture of crystalline polymorphs, including calcite and vaterite. On the other hand, guanidine and urea predominantly produce calcite. It is important to consider that the experimental observations are consistent with the original use of terms "chaotrope" and "kosmotrope," as previously discussed [3,69], and also that the chao-/kosmotorpic activities of salts and organic molecules can considerably deviate in solution mixtures. This discrepancy can originate from the non-additive effects of salts and organic molecules in inducing a net kosmotropic or chaotropic effect. For instance, simulated martian brine solutions are predominantly kosmotropic despite being constituted with chaotropic salts such as chlorides of iron and magnesium [70].

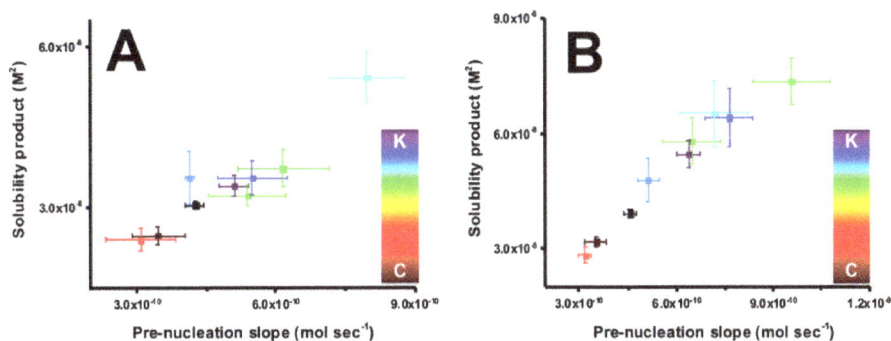

Figure 3. Scatter plots representing slopes of the pre-nucleation regime and corresponding solubility products of phases formed after nucleation at (**A**) pH 9.0 and (**B**) 9.75 in reference (black) and additive-containing (colored) titrations. The color scale represents the relative chao(C)-/kosmotropic(K) activities of guanidine, urea, mannitol, sorbitol, trehalose, glycine, betaine, and ectoine as determined by systematic studies on agar gelation [69]. Error bars depict ± one standard deviation.

2.2. On Sugar Stereoisomers

In order to examine possible contributions from a model that proposes solute-induced perturbations of the bulk structure of water, we investigate the effects of four sugar enantiomer pairs. The model suggests that liquid water consists of rapidly exchanging high and low density micro-domains, and the equilibrium between these domains can be altered by dissolved solutes [18,55]. This is supported by distinct elution profiles for glucose enantiomers, suggested to involve micro-domain equilibrium shifts, wherein the bioactive enantiomer prefers a less dense aqueous environment and L-glucose favors a more dense water state [55]. Based on this hypothesis, the individual enantiomers are expected to have distinct effects on nucleation independent of solution pH. However, at pH 9.75, L-glucose stabilizes PNCs and leads to thermodynamically more stable nucleation products (Figure 4). In relation, D-glucose induces no significant effects on the nucleation of CaCO$_3$ particles. At lower pH, the effects of L-glucose are minor relative to those at pH 9.75. Under kinetically controlled mineralization, both enantiomers produce calcite as the predominant mineral polymorph (Figure S2). In light of these results, a pH-independent perturbation to the bulk solvent structure by the enantiomeric additives is not validated or is too weak and eclipsed by direct interactions to have observable effects on mineral nucleation.

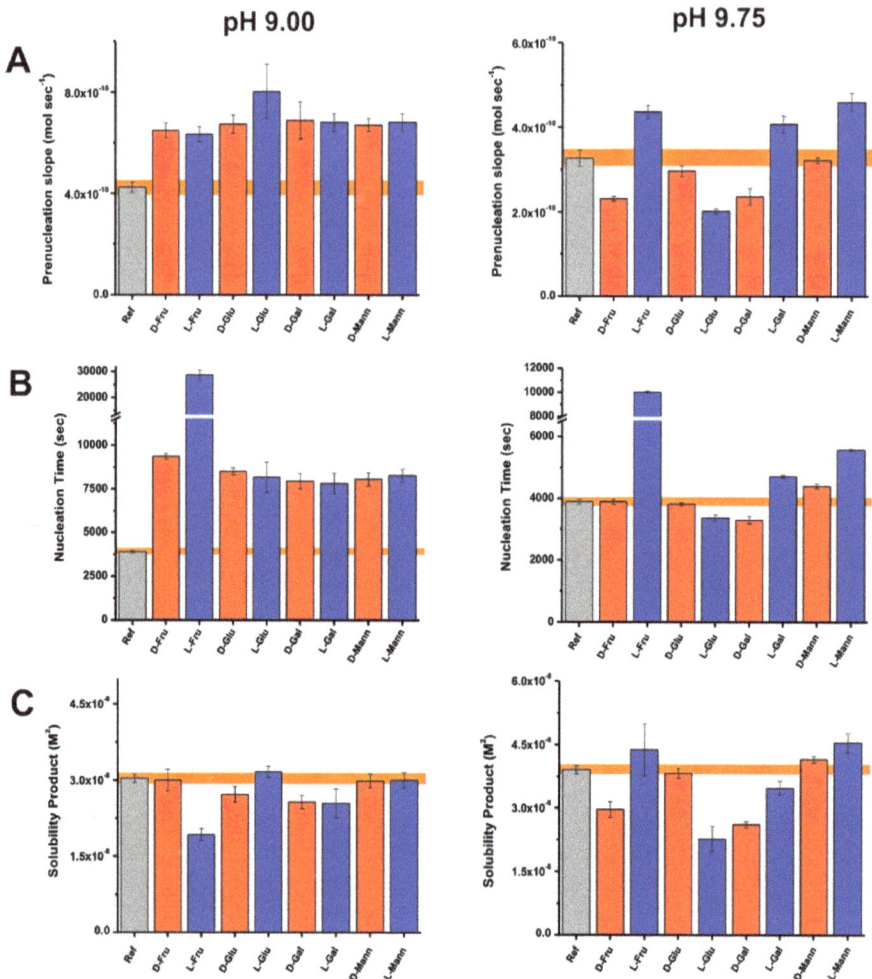

Figure 4. Bar plots representing the early stages of $CaCO_3$ nucleation in terms of the (**A**) pre-nucleation slope, (**B**) time required for nucleation, and (**C**) solubility of initially nucleated phase at pH 9.0 (left) and 9.75 (right) for additive-containing and reference (Ref) experiments. The acronyms indicate fructose (Fru), glucose (Glu), galactose (Gal), and mannose (Mann) in D (red) and L (blue) forms. Error bars depict ± one standard deviation. Error values for reference experiments are shaded (orange).

Similar trends are noted for three sugar enantiomers, namely fructose, galactose and mannose. For these additives, the distinct effects of the D and L isomers on mineralization at pH 9.75 are relatively nullified at pH 9.0 (Figure 4). For instance, D-galactose and D-fructose stabilize PNCs and induce nucleation products with lower solubilities relative to the reference experiments, specifically at pH 9.75. Therefore, the effects of sugars on mineralization cannot be absolutely associated with either the D or L enantiomeric forms but are also determined by pH-dependent parameters such as the $CO_3^{2-}:HCO_3^-$ ratio and the additive conformation. Conditions of pH can alter the chemical nature of ion-clusters and the subsequent interactions between additives and mineral precursors [68,71]. Therefore, the selective trends of mineral nucleation are likely due to direct interactions with mineral species and associated hydration rather than perturbations to the bulk water phase.

3. Discussion

This study identifies the distinct trends of the early stages of precipitation of calcium carbonate in the presence of small molecules, classified on their chao/kosmotropic nature and stereochemistry. Bearing in mind the similar size regimes but distinct physicochemistry of ion-associates and macromolecules, the following mechanisms are proposed for the additive-controlled nucleation process. First, the destabilizers of ion-association, such as betaine and ectoine, are possibly excluded from the hydration shells of ion-clusters. Since ion association of CaCO$_3$ PNCs is driven by the release of hydration waters [29], stabilized hydration shells would have adverse effects. Indeed, molecular dynamics simulations reveal that ectoine is preferentially excluded from macromolecular surfaces, leading to a dense hydration layer [72]. During the early stages of mineralization, similar interactions might suppress dehydration processes related to ion-association and their aggregation/coalescence toward mineral precursors with a larger degree of structural order (Figure 5A). Certain polyalcohols such as mannitol and sorbitol also destabilize PNCs and lead to more soluble nucleation products, relative to the reference experiments. The polyalcohols lead to pronounced destabilization of PNCs in comparison to the sugars, possibly due to the conformational flexibility originating from the lack of a constraining ring structure. Substantial increases in the surface tension of water by polyalcohols also lead to a preferential hydration of biomolecular co-solutes [73,74]. Similar interactions might apply to the formation of PNCs and other transient mineral phases in the presence of polyalcohols, during which dehydration reactions underlying the maturation of mineral precursors are inhibited to a certain extent. Among the kosmotropes tested, trehalose is the most potent in suppressing ion-association. A possible explanation is the property of trehalose molecules in effectively forming hydrogen bonds with water molecules distributed homogenously in the solvent [75,76]. Therefore, in addition to increased surface tension of the solvent, the ion species weakly associated with trehalose are possibly also broadly distributed in the solvent and inhibited against ion-association.

An interesting observation is the promotion of nucleation (F < 1.0) in the presence of certain kosmotropes and sugars. The observed effects might be due to an interference with the liquid-liquid demixing and dehydration processes involved in particle nucleation and affecting its kinetics [33,77–79]. The free energy change of solvation involves contributions from the formation of the cavity, which the solute occupies, and solute-solvent interactions (van der Waals and electrostatic) [80], expressed as follows:

$$\delta(\Delta G^0)_{\text{solvation}} = \delta(\Delta G^0)_{\text{cavitation}} + \delta(\Delta G^0)_{\text{vW}} + \delta(\Delta G^0)_{\text{elec}}$$

This equation shows that kosmotropic additives that generally lower the solubility of solutes (salting out) suppress the formation of solvent cavities (first term in the above equation), and/or unfavorably affect the solute-solvent interactions (second and third term). Either effect could render liquid-like mineral precursors relatively more transient in presence of these additives by promoting the dehydration of inorganic liquid-like intermediate, thereby accelerating the nucleation of solid particles in a minor but occasionally detectable manner (this observation is not general for kosmotropes).

Considering the stabilizing effects of chaotropes toward PNCs, their influence is likely due to direct interactions with the inorganic species (Figure 5B). Previous investigations show that urea binds to the hydrophilic surfaces of proteins and stabilizes non-native conformations by usurping hydrogen bonds with hydrophilic surfaces [81]. The basis for this surfactant-like action of chaotropes is enthalpy driven and related to changes in interfacial energies and hydrophobic interactions [81–84]. On similar lines, as polar nonelectrolytes, urea, thiourea and guanidine possibly capture the PNCs and serve as a "molecular glue" between the ion-clusters. This proposed mechanism is supported by the crystal structure of an urea derivative, wherein hydrogen-bond interactions with carbonate species involve two NH···O bonds [85]. Such interactions can potentially promote ion-association and shift the equilibrium towards stable clusters, subsequently yielding nucleation products that are thermodynamically more stable (Figure 2).

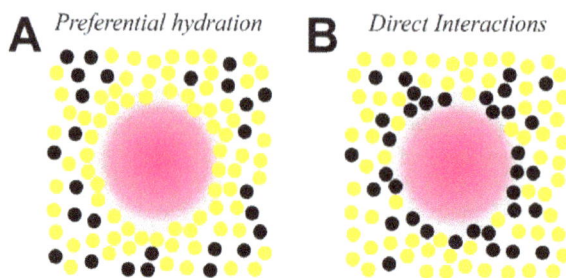

Figure 5. Schematic representations of the effects of the small organic molecules (black) and solvent molecules (yellow) on mineral precursors (pink), indicating (**A**) a preferential hydration of mineral co-solutes and (**B**) direct organic-inorganic interactions.

In view of these nucleation and mineralization studies (Figure 3, Figure S2), the accumulation of the compensatory kosmotropes such as trehalose and ectoine by halophilic microbes might serve as a means to suppress intracellular ion-association and mineral nucleation processes detrimental to survival of the organisms. Such effects may be enhanced by molecular crowding i.e., high cytosolic accumulation of these additives [86]. In this regard, the interactions between macromolecules and ions or ion-clusters are relevant toward defining the geochemical boundary conditions for life and predicting planetary habitability [69,87]. Such investigations indicate the non-additive nature of net chaotropic and kosmotrophic effects of solution mixtures of diverse salts and organic molecules. These might have crucial contributions during additive-controlled mineralization processes.

We also show that the role of chirality is not limited to crystal growth but also encompasses mineral nucleation. Previous studies show that calcite crystals grown in the presence of either L or D aspartic acid present asymmetry associated with the distribution of crystal steps and terraces [88,89], and even the CaCO$_3$ polymorph could be controlled by the chirality of an interacting amino acid [90]. The amplification of these interactions at macroscopic length scales is demonstrated by the distinct chirality of vaterite toroids and the helicity of potassium dichromate crystals tuned by the enantiospecificity of amino acids [91,92]. In view of mineral nucleation, we show that the sugar enantiomers present distinct interactions with mineral precursors and as a consequence affect the nucleation process in a pH-dependent manner. Thus with the distinct effects of amino acid and sugar enantiomers on mineral formation and growth, Lewis Carrol's allusion to the consequences of chirality "Perhaps looking-glass milk isn't good to drink" stands affirmed [93,94].

In summary, this systematic study explores the effects of small organic molecules on mineral nucleation. Additive-controlled mineralization is often represented as a two-component system involving the inorganic phase and the additive species, the solvent properties often being underrepresented. However, the contributions of solvophobic forces and weak interactions between the solvent molecules, additives, and material building units also appear crucial. It should be noted that PNCs are chemically distinct with highly dynamic configurations in comparison to the relative structural organization and rigidity of macromolecules. Therefore, future studies addressing the hydration dynamics during ion-association and phase transformation will certainly further elucidate the solvent phase as an active participant in nucleation and crystallization phenomena.

4. Materials and Methods

For the mineralization studies, the chemicals used are calcium chloride (1 M solution, Fluka, Happague, NY, USA), hydrochloric acid (1 M solution, Merck, Darmstadt, Germany,), sodium hydroxide (0.01 M, Alfa Aesar, Karlsruhe, Germany; 1 M, Merck, Darmstadt, Germany), sodium bicarbonate (Riedel de-Haën, Seelze, Germany), sodium carbonate (anhydrous, Sigma-Aldrich, Taufkirchen, Germany), sodium chloride (VWR Prolabo, Darmstadt, Germany), and ammonium carbonate (Acros Organics, Morris Plains, NJ, USA). The additives investigated include betaine (BioUltra, >99%, Sigma, Deisenhofen, Germany), ectoine (>99%, Sigma-Aldrich), D-(+)-trehalose from *Saccharomyces cerevisiae* (>99%, Sigma-Aldrich), D-sorbitol (>99%, Aldrich), D-mannitol (>98%, Sigma-Aldrich), glycine (>99%, Merck), urea (>99%, Merck), thiourea (>99%, Sigma-Aldrich), and guanidine (>99%, Sigma-Aldrich). The sugar additives used are D-(-)-fructose (Merck), D-galactose ($\geq$ 97%, Merck), α-D-(+)-glucose (Sigma-Aldrich, ACS reagent), D-mannose ($\geq$ 98%, Merck), L-(+)-fructose (>95%, Carbosynth, Compton, Berkshire, UK), L-(-)-galactose ($\geq$ 99%, Carbosynth), L-(-)-glucose ($\geq$ 99%, Sigma-Aldrich) and L-mannose ($\geq$ 98%, Merck).

Potentiometric titrations are performed with a commercial titration system from Metrohm (Filderstadt, Germany), in which an apparatus (Titrando 905) controls two dosing units (Dosino 800), operated by a customized software (Tiamo v2.2). A $CaCl_2$ (10 mM) solution is added at a fixed rate of 0.01 mL/min into 20 mL carbonate buffer (10 mM) containing additive (10 mM) [26,28,95]. Concurrently, the pH is maintained constant by the automatic counter-titration of NaOH (10 mM). The pH and free calcium contents are monitored by a glass electrode (Metrohm, No. 6.0256.100) and a Ca^{2+} ion-selective electrode (ISE, Metrohm, No. 6.0508.110). The ISE calibration is done by titrating $CaCl_2$ into 20 mL water at constant pH using the aforementioned methodology. All titrations were performed with stirring at 900 rpm and at room temperature with a minimum of three repetitions. The pH development and detected Ca^{2+} contents enable the quantitative evaluations of Ca^{2+} and CO_3^{2-} ions that occur in free and in bound states at a given time. Gas diffusion is performed by exposing solution mixtures containing $CaCl_2$ (10 mM) and additive (10 mM) to ammonium carbonate vapors in a previously described format [56].

Supplementary Materials: The following are available online at www.mdpi.com/2073-4352/7/10/302/s1, Figure S1: Simulated speciation of glycine in relation to pH conditions, Figure S2: Representative SEM images of $CaCO_3$ particles formed via gas diffusion experiments.

Acknowledgments: A.R. acknowledges fellowships from the Konstanz Research School Chemical Biology and Freiburg Institute for Advanced Studies. D.G. is a Research Fellow of the Zukunftskolleg of the University of Konstanz.

Author Contributions: Experimental tasks and preparation of the preliminary manuscript draft were done by A. R. Project supervision as well as detailed editing of manuscript were performed by D.G. and H.C.

Conflicts of Interest: The authors declare no conflict of interest.

References

1. Encrenaza, T.; Spohnb, T. Water in the solar system. In *Encyclopedia of Astrobiology*; Springer Berlin Heidelberg: Heidelberg, Germany, 2014.
2. Ball, P. Water as an active constituent in cell biology. *Chem. Rev.* **2008**, *108*, 74–108. [CrossRef] [PubMed]
3. Ball, P.; Hallsworth, J.E. Water structure and chaotropicity: Their uses, abuses and biological implications. *Phys. Chem. Chem. Phys.* **2015**, *17*, 8297–8305. [CrossRef] [PubMed]
4. Kunz, W.; Henle, J.; Ninham, B.W. 'Zur lehre von der wirkung der salze' (about the science of the effect of salts): Franz hofmeister's historical papers. *Curr. Opin. Colloid Inter.* **2004**, *9*, 19–37. [CrossRef]
5. Zhang, Y.; Furyk, S.; Bergbreiter, D.E.; Cremer, P.S. Specific ion effects on the water solubility of macromolecules: Pnipam and the hofmeister series. *J. Am. Chem. Soc.* **2005**, *127*, 14505–14510. [CrossRef] [PubMed]
6. Mason, P.E.; Dempsey, C.E.; Vrbka, L.; Heyda, J.; Brady, J.W.; Jungwirth, P. Specificity of ion-protein interactions: Complementary and competitive effects of tetrapropylammonium, guanidinium, sulfate, and chloride ions. *J. Phys. Chem. B* **2009**, *113*, 3227–3234. [CrossRef] [PubMed]
7. McCammick, E.M.; Gomase, V.S.; McGenity, T.J.; Timson, D.J.; Hallsworth, J.E. Water-hydrophobic compound interactions with the microbial cell. In *Handbook of Hydrocarbon and Lipid Microbiology*; Springer Berlin Heidelberg: Heidelberg, Germany, 2010.
8. Kunz, W. Specific ion effects in colloidal and biological systems. *Curr. Opin. Colloid Inter.* **2010**, *15*, 34–39. [CrossRef]

9. Vrbka, L.; Jungwirth, P.; Bauduin, P.; Touraud, D.; Kunz, W. Specific ion effects at protein surfaces: A molecular dynamics study of bovine pancreatic trypsin inhibitor and horseradish peroxidase in selected salt solutions. *J. Phys. Chem. B* **2006**, *110*, 7036–7043. [CrossRef] [PubMed]

10. Yancey, P.H.; Clark, M.E.; Hand, S.C.; Bowlus, R.D.; Somero, G.N. Living with water stress: Evolution of osmolyte systems. *Science* **1982**, *217*, 1214–1222. [CrossRef] [PubMed]

11. Lever, M.; Blunt, J.W.; Maclagan, R.G.A.R. Some ways of looking at compensatory kosmotropes and different water environments. *Comp. Biochem. Physiol. A* **2001**, *130*, 471–486. [CrossRef]

12. Crowe, J.; Crowe, L. Membrane integrity in anhydrobiotic organisms: Toward a mechanism for stabilizing dry cells. In *Water and Life*; Springer: Berlin, Germany, 1992; pp. 87–103.

13. Albertyn, J.; Hohmann, S.; Thevelein, J.M.; Prior, B.A. GPD1, which encodes glycerol-3-phosphate dehydrogenase, is essential for growth under osmotic stress in saccharomyces cerevisiae, and its expression is regulated by the high-osmolarity glycerol response pathway. *Mol. Cell. Biol.* **1994**, *14*, 4135–4144. [CrossRef] [PubMed]

14. Gekko, K.; Timasheff, S.N. Mechanism of protein stabilization by glycerol: Preferential hydration in glycerol-water mixtures. *Biochemistry* **1981**, *20*, 4667–4676. [CrossRef] [PubMed]

15. Courtenay, E.S.; Capp, M.W.; Anderson, C.F.; Record, M.T. Vapor pressure osmometry studies of osmolyte-protein interactions: Implications for the action of osmoprotectants in vivo and for the interpretation of "osmotic stress" experiments in vitro. *Biochemistry* **2000**, *39*, 4455–4471. [CrossRef] [PubMed]

16. Moelbert, S.; Normand, B.; De Los Rios, P. Kosmotropes and chaotropes: Modelling preferential exclusion, binding and aggregate stability. *Biophy. Chem.* **2004**, *112*, 45–57. [CrossRef] [PubMed]

17. Bolen, D.W.; Baskakov, I.V. The osmophobic effect: Natural selection of a thermodynamic force in protein folding. *J. Mol. Biol.* **2001**, *310*, 955–963. [CrossRef] [PubMed]

18. Wiggins, P.M. High and low density intracellular water. *Cell. Mol. Biol.* **2001**, *47*, 735–744. [PubMed]

19. Nozaki, Y.; Tanford, C. The solubility of amino acids, diglycine, and triglycine in aqueous guanidine hydrochloride solutions. *J. Biol. Chem.* **1970**, *245*, 1648–1652. [PubMed]

20. De Xammar Oro, J. Role of co-solute in biomolecular stability: Glucose, urea and the water structure. *J. Biol. Phys.* **2001**, *27*, 73–79. [CrossRef] [PubMed]

21. Timasheff, S.N. Protein-solvent preferential interactions, protein hydration, and the modulation of biochemical reactions by solvent components. *Proc. Natl. Acad. Sci. USA* **2002**, *99*, 9721–9726. [CrossRef] [PubMed]

22. Chin, J.P.; Megaw, J.; Magill, C.L.; Nowotarski, K.; Williams, J.P.; Bhaganna, P.; Linton, K.; Patterson, M.F.; Underwood, G.J.C.; Mswaka, A.Y.; et al. Solutes determine the temperature windows for microbial survival and growth. *Proc. Natl. Acad. Sci. USA* **2010**, *107*, 7835–7840. [CrossRef] [PubMed]

23. Batchelor, J.D.; Olteanu, A.; Tripathy, A.; Pielak, G.J. Impact of protein denaturants and stabilizers on water structure. *J. Am. Chem. Soc.* **2004**, *126*, 1958–1961. [CrossRef] [PubMed]

24. Rothon, R.; Paynter, C. Calcium carbonate fillers. In *Encyclopedia of Polymers and Composites*; Palsule, S., Ed.; Springer Berlin Heidelberg: Heidelberg, Germany, 2015; pp. 1–9.

25. Verch, A.; Gebauer, D.; Antonietti, M.; Cölfen, H. How to control the scaling of CaCO$_3$: A "fingerprinting technique" to classify additives. *Phys. Chem. Chem. Phys.* **2011**, *13*, 16811–16820. [CrossRef] [PubMed]

26. Gebauer, D.; Cölfen, H.; Verch, A.; Antonietti, M. The multiple roles of additives in CaCO$_3$ crystallization: A quantitative case study. *Adv. Mat.* **2009**, *21*, 435–439. [CrossRef]

27. Demichelis, R.; Raiteri, P.; Gale, J.D.; Quigley, D.; Gebauer, D. Stable prenucleation mineral clusters are liquid-like ionic polymers. *Nat. Comm.* **2011**, *2*, 590. [CrossRef] [PubMed]

28. Gebauer, D.; Völkel, A.; Cölfen, H. Stable prenucleation calcium carbonate clusters. *Science* **2008**, *322*, 1819–1822. [CrossRef] [PubMed]

29. Kellermeier, M.; Raiteri, P.; Berg, J.K.; Kempter, A.; Gale, J.D.; Gebauer, D. Entropy drives calcium carbonate ion association. *Chemphyschem* **2015**, *17*, 3535–3541. [CrossRef] [PubMed]

30. Gehrke, N.; Cölfen, H.; Pinna, N.; Antonietti, M.; Nassif, N. Superstructures of calcium carbonate crystals by oriented attachment. *Cryst. Growth Des.* **2005**, *5*, 1317–1319. [CrossRef]

31. Penn, R.L.; Banfield, J.F. Imperfect oriented attachment: Dislocation generation in defect-free nanocrystals. *Science* **1998**, *281*, 969–971. [CrossRef] [PubMed]

32. De Yoreo, J.J.; Gilbert, P.U.; Sommerdijk, N.A.; Penn, R.L.; Whitelam, S.; Joester, D.; Zhang, H.; Rimer, J.D.; Navrotsky, A.; Banfield, J.F. Crystallization by particle attachment in synthetic, biogenic, and geologic environments. *Science* **2015**, *349*, aaa6760. [CrossRef] [PubMed]
33. Sebastiani, F.; Wolf, S.L.P.; Born, B.; Luong, T.Q.; Cölfen, H.; Gebauer, D.; Havenith, M. Water dynamics from THz Spectroscopy reveal the locus of a liquid-liquid binodal limit in aqueous $CaCO_3$ solutions. *Angew. Chem. Int. Ed.* **2017**, *56*, 490–495.
34. Gower, L.B. Biomimetic model systems for investigating the amorphous precursor pathway and its role in biomineralization. *Chem. Rev.* **2008**, *108*, 4551–4627. [CrossRef]
35. Wallace, A.F.; Hedges, L.O.; Fernandez-Martinez, A.; Raiteri, P.; Gale, J.D.; Waychunas, G.A.; Whitelam, S.; Banfield, J.F.; De Yoreo, J.J. Microscopic evidence for liquid-liquid separation in supersaturated $CaCO_3$ solutions. *Science* **2013**, *341*, 885–889. [CrossRef] [PubMed]
36. Burgos-Cara, A.; Putnis, C.V.; Rodriguez-Navarro, C.; Ruiz-Agudo, E. Hydration effects on the stability of calcium carbonate pre-nucleation species. *Minerals* **2017**, *7*, 126. [CrossRef]
37. Gebauer, D.; Cölfen, H. Prenucleation clusters and non-classical nucleation. *Nano. Today* **2011**, *6*, 564–584. [CrossRef]
38. Gebauer, D.; Gunawidjaja, P.N.; Ko, J.; Bacsik, Z.; Aziz, B.; Liu, L.; Hu, Y.; Bergström, L.; Tai, C.W.; Sham, T.K. Proto-calcite and proto-vaterite in amorphous calcium carbonates. *Angew. Chem.* **2010**, *49*, 8889–8891. [CrossRef] [PubMed]
39. Addadi, L.; Raz, S.; Weiner, S. Taking advantage of disorder: Amorphous calcium carbonate and its roles in biomineralization. *Adv. Mat.* **2003**, *15*, 959–970. [CrossRef]
40. Politi, Y.; Arad, T.; Klein, E.; Weiner, S.; Addadi, L. Sea urchin spine calcite forms via a transient amorphous calcium carbonate phase. *Science* **2004**, *306*, 1161–1164. [CrossRef] [PubMed]
41. Cartwright, J.H.; Checa, A.G.; Gale, J.D.; Gebauer, D.; Sainz-Díaz, C.I. Calcium carbonate polymorphism and its role in biomineralization: How many amorphous calcium carbonates are there? *Angew. Chem. Inter. Ed.* **2012**, *51*, 11960–11970. [CrossRef] [PubMed]
42. Radha, A.; Forbes, T.Z.; Killian, C.E.; Gilbert, P.; Navrotsky, A. Transformation and crystallization energetics of synthetic and biogenic amorphous calcium carbonate. *Proc. Natl. Acad. Sci. USA* **2010**, *107*, 16438–16443. [CrossRef] [PubMed]
43. Ihli, J.; Wong, W.C.; Noel, E.H.; Kim, Y.Y.; Kulak, A.N.; Christenson, H.K.; Duer, M.J.; Meldrum, F.C. Dehydration and crystallization of amorphous calcium carbonate in solution and in air. *Nat. Comm.* **2014**, *5*, 3169. [CrossRef] [PubMed]
44. Xu, X.-R.; Cai, A.-H.; Liu, R.; Pan, H.-H.; Tang, R.-K.; Cho, K. The roles of water and polyelectrolytes in the phase transformation of amorphous calcium carbonate. *J. Cryst. Growth* **2008**, *310*, 3779–3787. [CrossRef]
45. Luo, Y.; Sonnenberg, L.; Cölfen, H. Novel method for generation of additive free high-energy crystal faces and their reconstruction in solution. *Cryst. Growth Des.* **2008**, *8*, 2049–2051. [CrossRef]
46. Merzel, F.; Smith, J.C. Is the first hydration shell of lysozyme of higher density than bulk water? *Proc. Natl. Acad. Sci. USA* **2002**, *99*, 5378–5383. [CrossRef] [PubMed]
47. Pizzitutti, F.; Marchi, M.; Sterpone, F.; Rossky, P.J. How protein surfaces induce anomalous dynamics of hydration water. *J. Phys. Chem. B* **2007**, *111*, 7584–7590. [CrossRef] [PubMed]
48. Raiteri, P.; Gale, J.D. Water is the key to nonclassical nucleation of amorphous calcium carbonate. *J. Am. Chem. Soc.* **2010**, *132*, 17623–17634. [CrossRef] [PubMed]
49. Tritschler, U.; Kellermeier, M.; Debus, C.; Kempter, A.; Cölfen, H. A simple strategy for the synthesis of well-defined bassanite nanorods. *CrystEngComm* **2015**, *17*, 3772–3776. [CrossRef]
50. Khoshkhoo, S.; Anwar, J. Crystallization of polymorphs: The effect of solvent. *J. Phys. D* **1993**, *26*, B90. [CrossRef]
51. Niederberger, M.; Cölfen, H. Oriented attachment and mesocrystals: Non-classical crystallization mechanisms based on nanoparticle assembly. *Phys. Chem. Chem. Phys.* **2006**, *8*, 3271–3287. [CrossRef] [PubMed]
52. Polleux, J.; Pinna, N.; Antonietti, M.; Hess, C.; Wild, U.; Schlögl, R.; Niederberger, M. Ligand functionality as a versatile tool to control the assembly behavior of preformed titania nanocrystals. *Chem. Eur. J.* **2005**, *11*, 3541–3551. [CrossRef] [PubMed]
53. Zhang, H.; Banfield, J.F. Interatomic coulombic interactions as the driving force for oriented attachment. *CrystEngComm* **2014**, *16*, 1568–1578. [CrossRef]

54. Fichthorn, K.A. Atomic-scale aspects of oriented attachment. *Chem. Eng. Sci.* **2015**, *121*, 10–15. [CrossRef]

55. Wiggins, P. Life depends upon two kinds of water. *PLoS ONE* **2008**, *3*, e1406. [CrossRef] [PubMed]

56. Picker, A.; Kellermeier, M.; Seto, J.; Gebauer, D.; Cölfen, H. The multiple effects of amino acids on the early stages of calcium carbonate crystallization. *Z. Krist. Cryst. Mat.* **2012**, *227*, 744–757. [CrossRef]

57. Rao, A.; Berg, J.K.; Kellermeier, M.; Gebauer, D. Sweet on biomineralization: Effects of carbohydrates on the early stages of calcium carbonate crystallization. *Eur. J. Min.* **2014**, *26*, 537–552. [CrossRef]

58. Rao, A.; Fernández, M.S.; Cölfen, H.; Arias, J.L. Distinct effects of avian egg derived anionic proteoglycans on the early stages of calcium carbonate mineralization. *Cryst. Growth Des.* **2015**, *15*, 2052–2056. [CrossRef]

59. Rao, A.; Seto, J.; Berg, J.K.; Kreft, S.G.; Scheffner, M.; Cölfen, H. Roles of larval sea urchin spicule SM50 domains in organic matrix self-assembly and calcium carbonate mineralization. *J. Struc. Biol.* **2013**, *183*, 205–215. [CrossRef] [PubMed]

60. Gebauer, D.; Verch, A.; Borner, H.G.; Cölfen, H. Influence of selected artificial peptides on calcium carbonate precipitation—A quantitative study. *Cryst. Growth Des.* **2009**, *9*, 2398–2403. [CrossRef]

61. Kellermeier, M.; Gebauer, D.; Melero-García, E.; Drechsler, M.; Talmon, Y.; Kienle, L.; Cölfen, H.; García-Ruiz, J.M.; Kunz, W. Colloidal stabilization of calcium carbonate prenucleation clusters with silica. *Adv. Func. Mat.* **2012**, *22*, 4301–4311. [CrossRef]

62. Kellermeier, M.; Picker, A.; Kempter, A.; Cölfen, H.; Gebauer, D. A straightforward treatment of activity in aqueous CaCO₃ solutions and the consequences for nucleation theory. *Adv. Mat.* **2014**, *26*, 752–757. [CrossRef] [PubMed]

63. Wolf, S.L.; Jähme, K.; Gebauer, D. Synergy of mg 2+ and poly (aspartic acid) in additive-controlled calcium carbonate precipitation. *CrystEngComm* **2015**, *17*, 6857–6862. [CrossRef]

64. Verch, A.; Antonietti, M.; Cölfen, H. Mixed calcium-magnesium pre-nucleation clusters enrich calcium. *Z. Krist. Cryst. Mat.* **2012**, *227*, 718–722. [CrossRef]

65. De Lima Alves, F.; Stevenson, A.; Baxter, E.; Gillion, J.L.; Hejazi, F.; Hayes, S.; Morrison, I.E.; Prior, B.A.; McGenity, T.J.; Rangel, D.E.; et al. Concomitant osmotic and chaotropicity-induced stresses in Aspergillus wentii: Compatible solutes determine the biotic window. *Curr. Genetics* **2015**, *61*, 457–477. [CrossRef] [PubMed]

66. Zahn, D. Thermodynamics and kinetics of prenucleation clusters, classical and non-classical nucleation. *Chemphyschem* **2015**, *16*, 2069–2075. [CrossRef] [PubMed]

67. Gebauer, D.; Kellermeier, M.; Gale, J.D.; Bergstrom, L.; Cölfen, H. Pre-nucleation clusters as solute precursors in crystallisation. *Chem. Soc. Rev.* **2014**, *43*, 2348–2371. [CrossRef] [PubMed]

68. Rao, A.; Vásquez-Quitral, P.; Fernández, M.S.; Berg, J.K.; Sánchez, M.; Drechsler, M.; Neira-Carrillo, A.; Arias, J.L.; Gebauer, D.; Cölfen, H. Ph-dependent schemes of calcium carbonate formation in the presence of alginates. *Cryst. Growth Des.* **2016**, *16*, 1349–1359. [CrossRef]

69. Cray, J.A.; Russell, J.T.; Timson, D.J.; Singhal, R.S.; Hallsworth, J.E. A universal measure of chaotropicity and kosmotropicity. *Env. Microbiol.* **2013**, *15*, 287–296. [CrossRef] [PubMed]

70. Fox-Powell, M.G.; Hallsworth, J.E.; Cousins, C.R.; Cockell, C.S. Ionic strength is a barrier to the habitability of Mars. *Astrobiology* **2016**, *16*, 427–442. [CrossRef]

71. Rao, A.; Huang, Y.C.; Cölfen, H. Additive speciation and phase behavior modulate mineralization. *J. Phys. Chem. C* **2017**, doi:10.1021/acs.jpcc.7b02635. [CrossRef]

72. Yu, I.; Jindo, Y.; Nagaoka, M. Microscopic understanding of preferential exclusion of compatible solute ectoine: Direct interaction and hydration alteration. *J. Phys. Chem. B* **2007**, *111*, 10231–10238. [CrossRef] [PubMed]

73. Kaushik, J.K.; Bhat, R. Thermal stability of proteins in aqueous polyol solutions: Role of the surface tension of water in the stabilizing effect of polyols. *J. Phys. Chem. B* **1998**, *102*, 7058–7066. [CrossRef]

74. Kaushik, J.K.; Bhat, R. Why is trehalose an exceptional protein stabilizer? An analysis of the thermal stability of proteins in the presence of the compatible osmolyte trehalose. *J. Biol. Chem.* **2003**, *278*, 26458–26465. [CrossRef] [PubMed]

75. Lerbret, A.; Bordat, P.; Affouard, F.; Descamps, M.; Migliardo, F. How homogeneous are the trehalose, maltose, and sucrose water solutions? An insight from molecular dynamics simulations. *J. Phys. Chem. B* **2005**, *109*, 11046–11057. [CrossRef] [PubMed]

76. Jain, N.K.; Roy, I. Effect of trehalose on protein structure. *Prot. Sci.* **2009**, *18*, 24–36. [CrossRef] [PubMed]

77. Gutowski, K.E.; Broker, G.A.; Willauer, H.D.; Huddleston, J.G.; Swatloski, R.P.; Holbrey, J.D.; Rogers, R.D. Controlling the aqueous miscibility of ionic liquids: Aqueous biphasic systems of water-miscible ionic liquids and water-structuring salts for recycle, metathesis, and separations. *J. Am. Chem. Soc.* **2003**, *125*, 6632–6633. [CrossRef] [PubMed]

78. Abdolrahimi, S.; Nasernejad, B.; Pazuki, G. Influence of process variables on extraction of cefalexin in a novel biocompatible ionic liquid based-aqueous two phase system. *Phys. Chem. Chem. Phys.* **2015**, *17*, 655–669. [CrossRef] [PubMed]

79. Bewernitz, M.A.; Gebauer, D.; Long, J.; Cölfen, H.; Gower, L.B. A metastable liquid precursor phase of calcium carbonate and its interactions with polyaspartate. *Faraday Discuss.* **2012**, *159*, 291–312. [CrossRef]

80. Colominas, C.; Luque, F.J.; Teixidó, J.; Orozco, M. Cavitation contribution to the free energy of solvation: Comparison of different formalisms in the context of MST calculations. *Chem. Phys.* **1999**, *240*, 253–264. [CrossRef]

81. Bennion, B.J.; Daggett, V. The molecular basis for the chemical denaturation of proteins by urea. *Proc. Natl. Acad. Sci. USA* **2003**, *100*, 5142–5147. [CrossRef] [PubMed]

82. Wallqvist, A.; Covell, D.; Thirumalai, D. Hydrophobic interactions in aqueous urea solutions with implications for the mechanism of protein denaturation. *J. Am. Chem. Soc.* **1998**, *120*, 427–428. [CrossRef]

83. Zangi, R.; Zhou, R.; Berne, B.J. Urea's action on hydrophobic interactions. *J. Am. Chem. Soc.* **2009**, *131*, 1535–1541. [CrossRef] [PubMed]

84. England, J.L.; Pande, V.S.; Haran, G. Chemical denaturants inhibit the onset of dewetting. *J. Am. Chem. Soc.* **2008**, *130*, 11854–11855. [CrossRef] [PubMed]

85. Boiocchi, M.; Del Boca, L.; Gómez, D.E.; Fabbrizzi, L.; Licchelli, M.; Monzani, E. Nature of urea-fluoride interaction: Incipient and definitive proton transfer. *J. Am. Chem. Soc.* **2004**, *126*, 16507–16514. [CrossRef] [PubMed]

86. Rao, A.; Cölfen, H. On the biophysical regulation of mineral growth: Standing out from the crowd. *J. Struc. Biol.* **2016**, *196*, 232–243. [CrossRef] [PubMed]

87. Yakimov, M.M.; La Cono, V.; Spada, G.L.; Bortoluzzi, G.; Messina, E.; Smedile, F.; Arcadi, E.; Borghini, M.; Ferrer, M.; Schmitt-Kopplin, P.; et al. Microbial community of the deep-sea brine Lake Kryos seawater–brine interface is active below the chaotropicity limit of life as revealed by recovery of mRNA. *Env. Microbiol.* **2015**, *17*, 364–382. [CrossRef]

88. Addadi, L.; Weiner, S. Biomineralization: Crystals, asymmetry and life. *Nature* **2001**, *411*, 753–755. [CrossRef] [PubMed]

89. Orme, C.; Noy, A.; Wierzbicki, A.; McBride, M.; Grantham, M.; Teng, H.; Dove, P.; DeYoreo, J. Formation of chiral morphologies through selective binding of amino acids to calcite surface steps. *Nature* **2001**, *411*, 775–779. [CrossRef] [PubMed]

90. Wolf, S.E.; Loges, N.; Mathiasch, B.; Panthöfer, M.; Mey, I.; Janshoff, A.; Tremel, W. Phase selection of calcium carbonate through the chirality of adsorbed amino acids. *Angew. Chem. Int. Ed.* **2007**, *46*, 5618–5623. [CrossRef] [PubMed]

91. Oaki, Y.; Imai, H. Amplification of chirality from molecules into morphology of crystals through molecular recognition. *J. Am. Chem. Soc.* **2004**, *126*, 9271–9275. [CrossRef] [PubMed]

92. Jiang, W.; Pacella, M.S.; Athanasiadou, D.; Nelea, V.; Vali, H.; Hazen, R.M.; Gray, J.J.; McKee, M.D. Chiral acidic amino acids induce chiral hierarchical structure in calcium carbonate. *Nat. Comm.* **2017**, *8*. [CrossRef] [PubMed]

93. Barron, L. From cosmic chirality to protein structure and function: Lord Kelvin's legacy. *QJM* **1997**, *90*, 793–800. [CrossRef] [PubMed]

94. Heilbronner, E.; Dunitz, J.D. *Reflections on Symmetry in Chemistry and Elsewhere*; John Wiley & Sons: Hoboken, NJ, USA, 1993.

95. Kellermeier, M.; Cölfen, H.; Gebauer, D. Investigating the early stages of mineral precipitation by potentiometric titration and analytical ultracentrifugation. *Methods Enzymol.* **2013**, *532*, 45–69. [PubMed]

crystals

MDPI

Communication

Over-Production, Crystallization, and Preliminary X-ray Crystallographic Analysis of a Coiled-Coil Region in Human Pericentrin

Min Ye Kim, Jeong Kuk Park, Yeowon Sim, Doheum Kim, Jeong Yeon Sim and SangYoun Park *

School of Systems Biomedical Science, Soongsil University, Seoul 06978, Korea; abpmrrc@nate.com (M.Y.K.); water1028@naver.com (J.K.P.); rato0001@naver.com (Y.S.); dohuemi@naver.com (D.K.); sjyeon1103@naver.com (J.Y.S.)
* Correspondence: psy@ssu.ac.kr; Tel.: +82-2-820-0456

Academic Editor: Jolanta Prywer
Received: 12 September 2017; Accepted: 28 September 2017; Published: 2 October 2017

Abstract: The genes encoding three coiled-coil regions in human pericentrin were gene synthesized with *Escherichia coli* codon-optimization, and the proteins were successfully over-produced in large quantities using *E. coli* expression. After verifying that the purified proteins were mostly composed of α-helices, one of the proteins was crystallized using polyethylene glycol 8000 as crystallizing agent. X-ray diffraction data were collected to 3.8 Å resolution under cryo-condition using synchrotron X-ray. The crystal belonged to space group C2 with unit cell parameters a = 324.9 Å, b = 35.7 Å, c = 79.5 Å, and β = 101.6°. According to Matthews' coefficient, the asymmetric unit may contain up to 12 subunits of the monomeric protein, with a crystal volume per protein mass (V_M) of 1.96 $Å^3$ Da^{-1} and a 37.3% solvent content.

Keywords: pericentrin; coiled-coil; centrosome; pericentriolar material (PCM)

1. Introduction

The centrosome is the main microtubule organizing center (MTOC) in animal cells. In non-mitotic interphase cells, the centrosome is located near the nucleus to produce an assembly of microtubules that radiates towards the cell periphery, serving as tracks for motor protein-mediated transport of cellular compartments. During mitosis, microtubules reorganize from the centrosomes to form spindle poles that accurately segregate the duplicated chromosomes in two. The centrosome consists of a pair of orthogonally arranged centrioles surrounded by pericentriolar material (PCM). The PCM includes factors such as the γ-tubulin ring complex (γ-TuRC) that directly function to nucleate the microtubule arrays [1–4]. Pericentrin (PCNT) is a large ~360 kDa protein that also exists within PCM [5] to act as a scaffold for anchoring multiple proteins of the PCM [6]. PCNT contains a series of predicted coiled-coil regions over most of their length [5], but a highly conserved PCM targeting motif called the PACT domain is found near the C-terminus [7] (Figure 1). PCNT has been linked to many human disorders [6], and one of them is the loss-of-function mutations that cause microcephalic osteodysplastic primordial dwarfism type II (MOPD II) [8].

In this study, three regions in the human PCNT which are predicted as coiled-coils were successfully over-produced in *Escherichia coli* using plasmids containing *E. coli* codon-optimized PCNT gene. The high α-helical contents for the three PCNT proteins were further confirmed by circular dichroism analysis, and one of them was crystallized for preliminary X-ray crystallographic analysis.

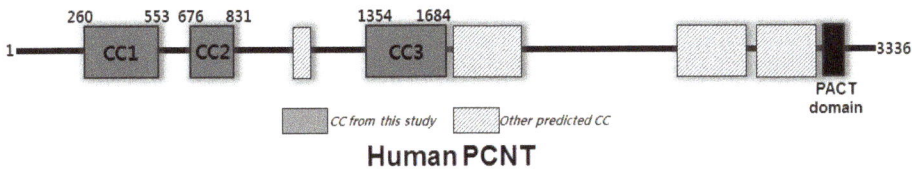

Human PCNT

Figure 1. Regions of human pericentrin (PCNT) showing predicted coiled-coil (CC) and the PACT domain.

2. Materials and Methods

2.1. Macromolecule Production

The DNA encoding three regions in human pericentrin (PCNT, full length of residues 1–3336) which are predicted to encode coiled-coil (CC) motifs (CC1, 260–553; CC2, 676–831; CC3, 1354–1684) were synthesized with an addition of nucleotides encoding N-terminal His-tag (Bioneer, Daejeon, Korea) for affinity purification. All genes were codon-optimized for expression in *E. coli* and cloned into pET28a vector (Merck, Kenilworth, NJ, USA) using *Nde*I and *Bam*HI restriction enzyme sites. The generated plasmids were all sequence verified of the insert region, and were used to transform the *E. coli* BL21 (DE3) (Merck, Kenilworth, NJ, USA) cells using heat shock at 42 °C (45 s). The transformed cells were grown at 37 °C in 1 L of Luria-Bertani (LB) medium to an OD_{600} of ~0.8 in the presence of 25 µg/mL kanamycin. Expression of the recombinant protein was induced by the addition of 0.5 mM isopropyl-D-thiogalactopyranoside (IPTG) at 22 °C, and cells were allowed to grow for an extra 16 h. Cells were harvested using centrifugation at $4500\times$ g for 10 min (4 °C). All three N-terminal His_6-tagged proteins of PCNT were over-produced with soluble expression of the proteins. For protein purification, the bacterial cell pellets were re-suspended in 50 mL ice-cold lysis buffer (20 mM Tris pH 7.5, 500 mM NaCl, and 5 mM imidazole) and lysed on ice by sonication. The homogenates were centrifuged at $70,000\times$ g for 30 min (4 °C), and supernatants poured over a 5 mL Ni-nitrilotriacetic acid agarose (Ni-NTA) (Qiagen, Hilden, Germany) gravity column. The columns were washed with five column volumes of wash buffer (20 mM Tris pH 7.5, 20 mM imidazole, and 500 mM NaCl), and the proteins were eluted with elution buffer (20 mM Tris pH 7.5, 200 mM imidazole, and 500 mM NaCl). The elution fractions containing the PCNT proteins were checked using Bradford assay (BioRad, Berkeley, CA, USA), combined, and added with 50 µL of 0.25 U/µL bovine thrombin (Invitrogen, Carlsbad, CA, USA). After proteolysis of the His_6-tag for 16 h incubation at 4 °C, the protein samples were further purified using a HiLoad® 26/60 Superdex® 200 size-exclusion column (SEC) pre-equilibrated with SEC buffer (50 mM Tris pH 7.5, 150 mM NaCl, and 2 mM DTT). The proteolysis mixes were loaded into the column connected to an ÄKTA FPLC system (GE Healthcare, Little Chalfont, UK). The elution profiles of all three proteins showed one major peak (Figure 2), and the fractions were concentrated by Amicon®ultracentrifugation filters (Merck). Final protein concentrations were estimated by A_{280} with a molar extinction coefficient (CC1, 15220 M^{-1} cm^{-1}; CC2, 6970 M^{-1} cm^{-1}; CC3, 11380 M^{-1} cm^{-1}) calculated based upon the numbers of tryptophan and tyrosine residues [9]. Purity and homogeneity were assessed using SDS-PAGE analysis (Figure 2). The concentrated proteins in SEC buffer were flash-cooled and stored in liquid nitrogen.

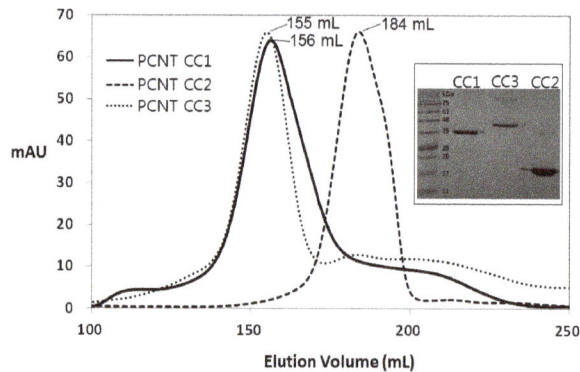

Figure 2. Size-exclusion chromatograms of three PCNT coiled-coil proteins (CC1, CC2, and CC3) and SDS-PAGE analysis of the concentrated fractions under the elution peaks (inset).

2.2. Circular Dichroism (CD) Studies for Secondary Structure Estimation

The contents of the secondary structure elements in the expressed PCNT proteins were estimated by scanning ellipticity over wavelength (200–240 nm) using a JASCO spectropolarimeter (Model J-810, Tokyo, Japan). The three PCNT proteins with the final concentration of 0.3 µg/mL were analyzed using a 0.1 cm path-length cuvette to obtain the spectra. Secondary structure estimation programs of K2D2 [10] and K2D3 [11] were applied to the CD data for approximation of the secondary structure content. Because the fitting of CD data for secondary structure contents are highly dependent on the exact concentration of the protein (which in our case was estimated from calculated molar extinction coefficient), up to two-fold differences in the protein concentrations were allowed during the fits.

2.3. Crystallization

Although all three PCNT proteins were screened using commercial Wizard® solutions (Molecular Dimensions) for conditions to obtain crystals, only crystals of PCNT CC2 were obtained (Table 1). The crystal drops were set by mixing equal 1 µL volumes of the reservoir solution (10% (*w/v*) PEG 8K 100 mM potassium phosphate monobasic/sodium phosphate dibasic pH 6.2, and 200 mM NaCl) and the PCNT CC2 concentrate (60 mg/mL) in SEC buffer, and were equilibrated against 500 µL of the reservoir solution. Several crystals were harvested, gel-electrophoresed, and their tryptic digestion fragments were analyzed using LC-MS/MS (Table 2). The crystals were transferred to a cryo-protectant solution of 30% glycerol supplemented to the reservoir solution, flash-cooled in liquid nitrogen for storage, and transported to a synchrotron facility for diffraction experiment.

Table 1. Macromolecule production information for PCNT CC2.

Source Organism	Homo Sapiens
DNA source	Synthesized DNA
Cloning sites	*NdeI* and *BamHI*
Cloning vector	pET28a
Expression vector	pET28a
Expression host	*E. coli* BL21 (DE3)
Complete amino acid sequence of the construct produced [1]	MGSSHHHHHHSSGLVPRGSHM-^{676}EHKVQ^{680} ^{681}LSLLQTELKEEIELLKIENRNLYGKLQHET^{710} ^{711}RLKDDLEKVKHNLIEDHQKELNNAKQKTEL^{740} ^{741}MKQEFQRKETDWKVMKEELQREAEEKLTLM^{770} ^{771}LLELREKAESEKQTIINKFELREAEMRQLQ^{800} ^{801}DQQAAQILDLERSLTEQQGRLQQLEQDLTSD^{831}

[1] Non-native His$_6$-tag and thrombin site are underlined.

Table 2. List of peptide fragments found from mass spectrometry analysis of the crystallized PCNT CC2.

Peptide Sequence [1]	MH+ (Da)
GSHM-[676]EHKVQLSLLQTELKEEIELLK[696]	2932.59
[679]VQLSLLQTELKEEIELLKIENR[700]	2638.52
[678]KVQLSLLQTELKEEIELLK[696]	2254.33
[683]LLQTELKEEIELLKIENR[700]	2211.26
[676]EHKVQLSLLQTELKEEIELLK[696]	2520.44
[682]SLLQTELKEEIELLKIENR[700]	2298.30
[681]LSLLQTELKEEIELLKIENR[700]	2411.38
[793]EAEMRQLQDQQAAQILDLER[812]	2385.19
[684]LQTELKEEIELLKIENR[700]	2098.18
[679]VQLSLLQTELKEEIELLK[696]	2126.23
[682]SLLQTELKEEIELLK[696]	1786.02
[685]QTELKEEIELLKIENR[700]	1985.09
[681]LSLLQTELKEEIELLK[696]	1899.11
[683]LLQTELKEEIELLK[696]	1698.99
[721]HNLIEDHQKELNNAK[735]	1802.92
[686]TELKEEIELLKIENR[700]	1857.03
[685]QTELKEEIELLK[696]	1472.82
[719]VKHNLIEDHQKELNNAK[735]	2030.08
[684]LQTELKEEIELLK[696]	1585.91
[778]AESEKQTIINKFELR[792]	1805.98
[798]QLQDQQAAQILDLER[812]	1768.92
[687]ELKEEIELLKIENR[700]	1755.99
[767]LTLMLLELREKAESEK[782]	1903.06
[688]LKEEIELLKIENR[700]	1626.95
[798]QLQDQQAAQILDLERSLTEQQGR[820]	2668.37
[757]EELQREAEEKLTLMLLELR[775]	2343.26
[762]EAEEKLTLMLLELR[775]	1687.93
[762]EAEEKLTLMLLELREK[777]	1945.07

[1] Peptide hits are listed based on the sorted cross correlation score (XCorr) starting from 5.0 to 10.8 [XCorr is defined by Proteome Discoverer 1.3 (Thermo Scientific, USA)].

2.4. Data Collection and Processing

X-ray diffraction data were collected at 100 K using a CCD detector (ADSC Quantum 315r) at beamline 5C of the Pohang Light Source (PLS, Pohang, Korea). The crystal was rotated through a total of 180° with 1.0° oscillation range per frame. Data were processed in space group C2 using HKL2000 [12] (Table 3).

Table 3. Data collection and processing.

Diffraction Source	Pohang Light Source (PLS 5C) (Pohang, Korea)
Wavelength (Å)	0.9795
Temperature (K)	100
Detector	ADSC Quantum 315r
Crystal–detector distance (mm)	450
Rotation range per image (°)	1
Total rotation range (°)	180
Exposure time per image (s)	1
Space group	C2
a, b, c (Å)	324.9, 35.7, 79.5
α, β, γ (°)	90.0, 101.6, 90.0
Mosaicity (°)	1.1
Resolution range (Å)	50.0–3.80 (3.87–3.80)[1]
Total No. of reflections	16,221
No. of unique reflections	9074
Completeness (%)	97.3 (89.7)
Redundancy	1.7 (1.6)
$\langle I/\sigma(I) \rangle$	18.1 (3.9)
R_{merge}	0.463 (0.120)
$R_{p.i.m.}$	0.081 (0.303)
CC1/2	(0.918)
Overall B factor from Wilson plot (Å^2)	78.7

[1] Values for the outer shell are given in parentheses.

3. Results and Discussion

Various coiled-coil prediction algorithms such as COILS [13] suggest multiple occurrences of coiled-coil regions in the large-sized (3336 amino acids) human PCNT (Figure 1). Three human PCNT constructs of three coiled-coil regions were generated by gene synthesis with *E. coli* codon-optimization for bacterial recombinant expression. The proteins were successfully over-produced in *E. coli* as a soluble protein and purified with an overall yield of >50 mg per 1 L of LB culture. The SEC elution profiles of the three coiled-coil regions on HiLoad® 26/60 Superdex® 200 showed a single major peak, and SDS-PAGE analysis of the concentrated proteins under the peaks indicated successful expression of the proteins (Figure 2). The estimated molecular masses of the proteins based on SDS-PAGE standard protein markers were ~35 kDa (CC1), ~17 kDa (CC2), and ~40 kDa (CC3), which were as expected from the calculated molecular masses of the protein (CC1, 35.5 kDa; CC2, 19.2 kDa; CC3, 38.7 kDa).

Predicted to form mostly coiled-coils, the secondary structure contents of PCNT were expected to be largely α-helical. Circular dichroism (CD) analyses of the purified PCNTs confirmed this prediction (Figure 3a–c). When the experimental ellipticity values determined at different wavelengths (200–240 nm) were used for secondary structure estimations, the results indicated that large parts of the proteins were mostly α-helical with <5% being β-strand (Figure 3). For instance, PCNT CC1 was estimated with 69% (K2D2) or 81% (K2D3) α-helicity. Additionally, PCNT CC2 was estimated with 76% (K2D2) or 89% (K2D3) α-helicity, and PCNT CC3 with 63% (K2D2) or 67% (K2D3) α-helicity. Hence, the overall trend indicated that the three PCNT proteins were made up largely of α-helices as predicted from the amino acid sequences.

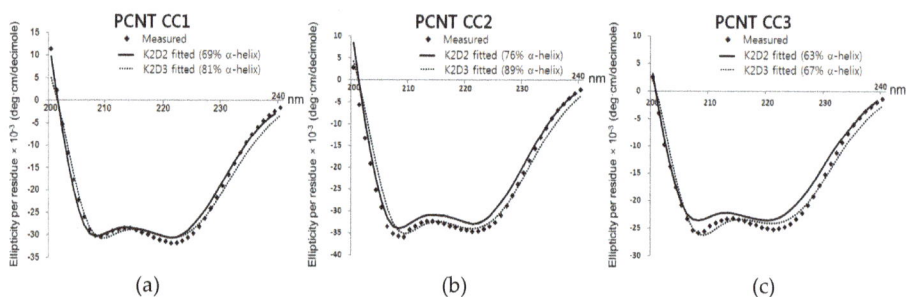

Figure 3. Circular dichroism studies for the estimation of secondary structure elements in PCNT (**a**) CC1, (**b**) CC2, and (**c**) CC3.

Among the three purified PCNT proteins, crystals were obtained only from PCNT CC2. Clustered crystals were grown against a reservoir solution of 10% (*w/v*) PEG 8K, 100 mM potassium phosphate monobasic/sodium phosphate dibasic pH 6.2, and 200 mM NaCl at 4 °C. Within two days, the crystals grew to approximately 5 μm × 50 μm × 100 μm (Figure 4), which was sufficient in size for X-ray diffraction experiments. The mass spectrometry analysis on the tryptic digestion fragments of the crystals confirmed the content of PCNT CC2 (Figure 4 and Table 2). The peptides found were mapped into the expressed PCNT CC2 protein sequence, resulting in 99.4% sequence coverage. Most crystals screened for X-ray diffraction using synchrotron radiation showed anisotropic diffraction. However, one out of ten crystals tested gave isotropic diffraction in all directions of oscillation to an average resolution limit of 3.8 Å (Figure 5). Alterations in PEG 8K concentrations, pH, as well as trials of different types of PEGs as the crystallization agent did not affect the overall diffraction quality. A total of 9074 unique reflections were measured and merged in the space group C2 (unit cell parameters of a = 324.9 Å, b = 35.7 Å, c = 79.5 Å, and β = 101.6°). The merged dataset was overall 97.3% complete with R_{merge} of 12% and $R_{p.i.m.}$ of 8.1% (50–3.80 Å). The statistics for the collected data are summarized in Table 3. According to the Matthews coefficient [14], the asymmetric unit may contain up to 12 subunits of the monomeric PCNT CC2 with a crystal volume per protein mass (V_M) of 1.96 Å^3 Da^{-1} and 37.3% solvent content. An asymmetric unit containing eight subunits is also plausible with V_M of 2.93 Å^3 Da^{-1} and 58.2% solvent content. A self-rotation function (kappa section = 180°) revealed a peak corresponding to two-fold noncrystallographic symmetry, further suggesting that PCNT CC2 exists as a dimer in the crystal. Because no model for PCNT exists in the Protein Databank (PDB), attempts for phasing via molecular replacement could not be made.

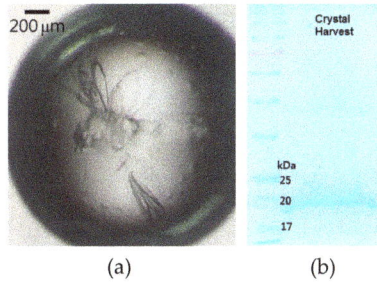

(a) (b)

Figure 4. (**a**) Clustered crystals of PCNT CC2 and (**b**) the SDS-PAGE analysis of harvested crystals used for MS analysis of tryptic digestion fragments.

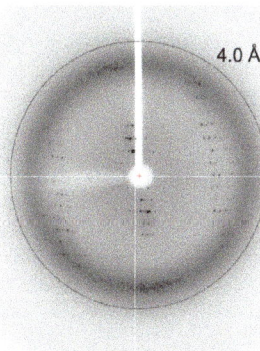

Figure 5. Representative diffraction image of the PCNT CC2 crystal.

In the future, we plan to determine the high-resolution structure of the coiled-coil PCNT CC2 by improving the current 3.8 Å resolution crystal and also by direct phasing. Since no information on the structure of PCNT exists, the structure would give insight towards understanding the mechanism that governs the function of PCNT.

Acknowledgments: The authors would like to thank the staff at PAL 5C beamline for their support and beam time. This research was supported by the Basic Science Program through the National Research Foundation of Korea (NRF) funded by the Ministry of Science, ICT & Future Planning (2016R1D1A1A09918187).

Author Contributions: Min Ye Kim and SangYoun Park conceived and designed the experiments; Min Ye Kim and Jeong Kuk Park performed the experiments; Jeong Kuk Park, Yeowon Sim, Doheum Kim and Jeong Yeon Sim analyzed the data; SangYoun Park wrote the paper.

Conflicts of Interest: The authors declare no conflict of interest.

References

1. Sluder, G. Two-way traffic: Centrosomes and the cell cycle. *Nat. Rev. Mol. Cell Biol.* **2005**, *6*, 743–748. [CrossRef] [PubMed]

2. Takahashi, M.; Yamagiwa, A.; Nishimura, T.; Mukai, H.; Ono, Y. Centrosomal proteins CG-NAP and kendrin provide microtubule nucleation sites by anchoring gamma-tubulin ring complex. *Mol. Biol. Cell* **2002**, *13*, 3235–3245. [CrossRef] [PubMed]
3. Zimmerman, W.C.; Sillibourne, J.; Rosa, J.; Doxsey, S.J. Mitosis-specific anchoring of gamma tubulin complexes by pericentrin controls spindle organization and mitotic entry. *Mol. Biol. Cell* **2004**, *15*, 3642–3657. [CrossRef] [PubMed]
4. Delaval, B.; Doxsey, S.J. Pericentrin in cellular function and disease. *J. Cell Biol.* **2010**, *188*, 181–190. [CrossRef] [PubMed]
5. Doxsey, S.; Zimmerman, W.; Mikule, K. Centrosome control of the cell cycle. *Trends Cell Biol.* **2005**, *15*, 303–311. [CrossRef] [PubMed]
6. Doxsey, S.J.; Stein, P.; Evans, L.; Calarco, P.D.; Kirschner, M. Pericentrin, a highly conserved centrosome protein involved in microtubule organization. *Cell* **1994**, *76*, 639–650. [CrossRef]
7. Gillingham, A.K.; Munro, S. The PACT domain, a conserved centrosomal targeting motif in the coiled-coil proteins AKAP450 and pericentrin. *EMBO Rep.* **2000**, *1*, 524–529. [CrossRef] [PubMed]
8. Rauch, A.; Thiel, C.T.; Schindler, D.; Wick, U.; Crow, Y.J.; Ekici, A.B.; van Essen, A.J.; Goecke, T.O.; Al-Gazali, L.; Chrzanowska, K.H.; et al. Mutations in the pericentrin (PCNT) gene cause primordial dwarfism. *Science* **2008**, *319*, 816–819. [CrossRef] [PubMed]
9. Gill, S.C.; von Hippel, P.H. Calculation of protein extinction coefficients from amino acid sequence data. *Anal. Biochem.* **1989**, *182*, 319–326. [CrossRef]
10. Perez-Iratxeta, C.; Andrade-Navarro, M.A. K2D2: Estimation of protein secondary structure from circular dichroism spectra. *BMC Struct. Biol.* **2008**, *8*, 1–5. [CrossRef] [PubMed]
11. Louis-Jeune, C.; Andrade-Navarro, M.A.; Perez-Iratxeta, C. Prediction of protein secondary structure from circular dichroism using theoretically derived spectra. *Proteins* **2012**, *80*, 374–381. [CrossRef] [PubMed]
12. Otwinowski, Z.; Minor, W. Processing of X-ray Diffraction Data Collected in Oscillation Mode. *Methods Enzymol.* **1997**, *276*, 307–326. [PubMed]
13. Lupas, A.; Van Dyke, M.; Stock, J. Predicting Coiled Coils from Protein Sequences. *Science* **1991**, *252*, 1162–1164. [CrossRef] [PubMed]
14. Matthews, B.W. Solvent content of protein crystals. *J. Mol. Biol.* **1968**, *33*, 491–497. [CrossRef]

crystals

MDPI

Article

Size and Shape Controlled Crystallization of Hemoglobin for Advanced Crystallography

Ayana Sato-Tomita and Naoya Shibayama *

Division of Biophysics, Department of Physiology, Jichi Medical University, 3311-1 Yakushiji, Shimotsuke, Tochigi 329-0498, Japan; ayana.sato@jichi.ac.jp
* Correspondence: shibayam@jichi.ac.jp; Tel.: +81-285-58-7308

Academic Editor: Jolanta Prywer
Received: 6 September 2017; Accepted: 17 September 2017; Published: 20 September 2017

Abstract: While high-throughput screening for protein crystallization conditions have rapidly evolved in the last few decades, it is also becoming increasingly necessary for the control of crystal size and shape as increasing diversity of protein crystallographic experiments. For example, X-ray crystallography (XRC) combined with photoexcitation and/or spectrophotometry requires optically thin but well diffracting crystals. By contrast, large-volume crystals are needed for weak signal experiments, such as neutron crystallography (NC) or recently developed X-ray fluorescent holography (XFH). In this article, we present, using hemoglobin as an example protein, some techniques for obtaining the crystals of controlled size, shape, and adequate quality. Furthermore, we describe a few case studies of applications of the optimized hemoglobin crystals for implementing the above mentioned crystallographic experiments, providing some hints and tips for the further progress of advanced protein crystallography.

Keywords: protein crystallization; hemoglobin; size control; shape control; macroseeding; X-ray crystallography; neutron crystallography; X-ray fluorescence holography; spectrophotometry

---

1. Introduction

Since the landmark work by Perutz and Kendrew on the X-ray structural determination of hemoglobin (Hb) and myoglobin (Mb) in the 1950s [1–3], X-ray crystallography (XRC) has been the most powerful technique to determine protein structures at the atomic level. In the early days of its history, however, it took several years of painstaking effort to solve a single protein structure due to the requirements of large, well-formed crystals (with edge dimensions of 0.5 mm or more) and heavy metal substitution for phase determination, and other experimental difficulties with data acquisition and analysis. Also, the identification of protein crystallization conditions was usually a trial-and-error process relied largely on the empirical knowledge of individual researchers. This is well illustrated by the fact that only eleven protein structures were solved by 1970. The situation has changed since the 1980s, with the advances in highly intense synchrotron X-ray sources, computer power, experimental instruments (such as X-ray detectors and automated sample manipulators), recombinant protein expression systems, phase determination techniques, and high-throughput/rationally-designed screening for crystallization conditions. Along with these technological and methodological advances, the crystal size and time required for XRC have been steadily decreasing over the years, and the initial search of crystallization conditions has become much easier and less time consuming. As a result, the number of crystal structures solved per year has been dramatically increasing from seven in 1980 to nearly ten thousand in 2016. Currently, the Protein Data Bank (PDB) [4] contains more than 120,000 protein structures, with the majority of them determined by XRC.

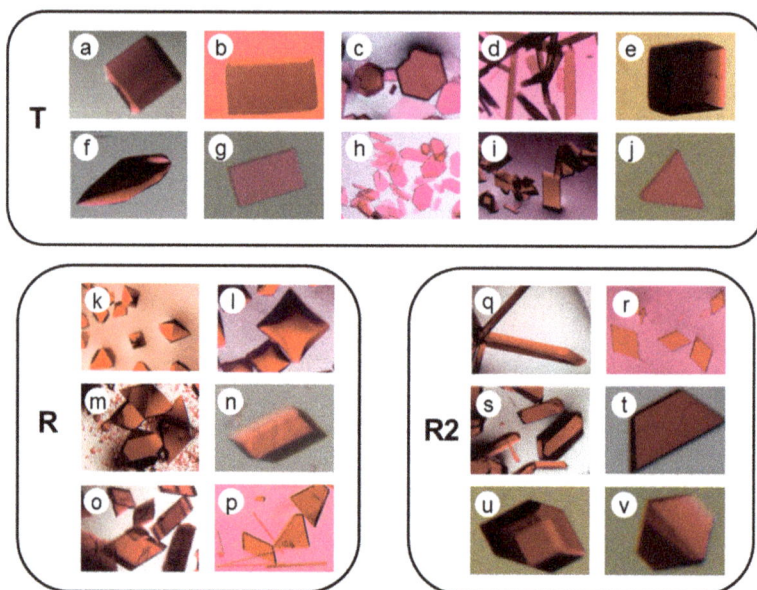

Figure 1. Examples of hemoglobin (Hb) crystals for three quaternary structures, T (**a**–**j**); R (**k**–**p**); and R2 (**q**–**v**); illustrating a variety of shapes and habits. (**a**) deoxyHbA with ammonium sulfate (pH 6.5); (**b**) deoxyHbA with PEG6000 (pH 7.0); (**c**) horse deoxyHb with PEG3350 (pH 8.8); (**d**) deoxyHbC with PEG3350 (pH 7.0); (**e**) deoxyHbC with PEG3350 (pH 8.6); (**f**) Ni(II)HbA with PEG3350 (pH 5.5); (**g**) Ni(II)HbS with PEG3350 (pH 8.3); (**h**) cross-linked Fe(II)-Ni(II) hybrid HbA with CO bound at the α hemes with PEG3350 (pH 6.6); (**i**) cross-linked Fe(II)-Ni(II) hybrid HbA with CO bound at the β hemes with PEG3350 (pH 6.6); (**j**) cross-linked Fe(II)-Ni(II) hybrid HbA with CO bound at only one β heme with PEG3350 (pH 7.0); (**k**) COHbA with phosphate (pH 6.7); (**l**) oxyHbA with ammonium sulfate (pH 7.1); (**m**) COHbC with phosphate (pH 7.2); (**n**) horse COHb with ammonium sulfate (pH 8.2); (**o**) horse deoxy desArg(α141)Hb with PEG3350 and PEG1000 (pH 7.7); (**p**) bezafibrate(BZF)-bound horse COHb with PEG1000 (pH 6.3); (**q**) COHbA with PEG3350 (pH 5.8, hanging drop); (**r**) COHbA with PEG3350 (pH 5.8, batch); (**s**) COHbC with PEG3350 (pH 7.6); (**t**) COHbS with PEG3350 (pH 8.3); (**u**) Fe(II)-Ni(II) hybrid HbS with CO bound at the α hemes with PEG3350 (pH 7.4); and (**v**) Fe(II)-Ni(II) hybrid HbS with CO bound at the β hemes with PEG3350 (pH 7.8). Note here that HbA is human adult hemoglobin, HbC and HbS are the naturally occurring mutant Hbs, in which Glu(β6) is replaced by Lys and Val, respectively, Ni(II)Hb is metal-substituted Hb in which Fe(II)-hemes have been replaced by Ni(II)-heme (acting as a surrogate for deoxy-Fe(II)-heme [5]), "cross-linked" denotes the presence of fumaryl cross-link between two Lys(β82) residues [6], and desArg(α141)Hb denotes Hb from which two Arg(α141) residues have been removed enzymatically.

Given such success, crystallization might be no longer such a severe bottleneck process in protein crystallography. However, this is not always the case, since protein crystallography experiments are becoming more diverse and challenging, and thereby requirements for crystals are also becoming more demanding in terms of size and shape. The purpose of this article is to illustrate our approach to the growth of Hb crystals to dimensions that meet the needs of advanced crystallography, including XRC combined with photoexcitation (e.g., photolysis of CO bound to Hb [7,8]) and/or spectrophotometry [9,10], neutron crystallography (NC) [11,12], and recently developed X-ray fluorescent holography (XFH) [13]. The first one requires optically thin but well-diffracting Hb crystals, while the latter two, especially XFH, require large-volume, well-ordered Hb crystals, due to feeble signals.

As described above, Hb was one of the first proteins whose crystal structure was solved by XRC, and so far hundreds of Hb structures have been solved and deposited in the PDB [4]. Human Hbs and other mammalian Hbs, together with avian, reptile, and bony fish Hbs, are all $(\alpha\beta)_2$ tetramers [14] that undergo oxygenation-linked quaternary structural change, characterized by a relative rotation between the two $\alpha\beta$ dimers [15]. Whereas, fully-unliganded deoxyHb tends to adopt a single tense (T) quaternary structure, fully-liganded Hb can adopt multiple relaxed quaternary structures, most typically represented by the classical relaxed (R) structure and the second relaxed (R2) structure (see a recent meta-analysis [16]). Thus, the tetrameric Hb structures can be classified into three distinct quaternary structures, T, R, and R2. An important fact to emphasize here is that each quaternary structure can be crystallized over a wide range of crystallization conditions and in different crystal lattices. This means that a number of different crystal forms are available by exploring various precipitates, crystallization conditions, and proteins with different amino acid sequences. Some examples from our laboratory are shown in Figure 1. Based on the accumulated experience and knowledge of Hb crystallization, we have chosen appropriate Hb samples and crystallization conditions for satisfying the requirements of the above mentioned crystallography experiments.

2. Results and Discussion

2.1. High-Quality, Thin Hb Crystal Growth for XRC Combined with Photoexcitation and Spectrophotometry

2.1.1. Human COHbC (β6 Glu→Lys) Crystal (R-State)

Even though XRC has provided atomic details of numerous protein structures, a mechanistic description of proteins also requires information about intermediates that occur during protein functioning. Hb and Mb can serve as model systems for studying such intermediates, by utilizing the photosensitivity of the CO bound complexes of these proteins (i.e., COHb and COMb). Absorption of photon breaks the heme Fe(II)-CO bond, initiating a series of structural and spectroscopic changes that can, in principle, be observed by XRC and spectrophotometry. Whereas, a number of crystal structures of photolyzed COMb have already been solved by several research groups [17–20], yet none of these groups have reported success in detailing the key movements in the Hb tetramer that initiate the allosteric transition. Since the quantum efficiency for CO photolysis is significantly lower for Hb than for Mb [21], the technical difficulties of producing strongly diffracting yet optically-thin crystals of COHb have prevented researchers from carrying out this type of experiment on Hb.

HbC is a human Hb variant in which the glutamic acid (Glu) at the sixth position of the β chain is replaced by lysine (Lys). It is known that this β6 Glu to Lys surface mutation does not significantly alter the oxygen equilibrium properties of Hb [22], but makes the protein less soluble when compared to normal human adult Hb (HbA) [23]. Indeed, HbC in the oxygenated form has a tendency to crystallize inside the red cell [24] (a typical example of pathological biogenic crystallization), which may contribute to "hemoglobin C disease" [23]. Recently we demonstrated that, although isomorphous with the COHbA crystals, the R-state COHbC crystals are of very high quality and remain intact upon deoxygenation [10], suggesting a high durability under photolysis conditions. We concluded that these COHbC crystals are well suited to the photolysis experiment, but unfortunately their shape is a problem.

The R-state COHbC crystals usually assume a tetragonal bipyramidal shape (Figure 2a), which is disadvantageous for CO photolysis and optical measurement. We therefore employed an approach of spatially restricted crystallization to change their shape from bipyramidal to a plate-like one (Figure 2b). First, a CO-saturated mixed solution of 2 µL of deionized 6.0% (w/v) COHbC and the same volume of 2.5 M phosphate buffer containing 0.80 M NaH_2PO_4 and 1.7 M K_2HPO_4 (pH 7.1) was spread out on a hole slide glass. After 10 s, the hole slide glass was covered by a standard slide glass with insertion of a holed silicone rubber to avoid drying out the crystallization solution (Figure 2c; right). Then, the sample was placed in a gas barrier bag and sealed with an oxygen absorber A-500HS (ISO, Yokohama, Japan) and an oxygen indicator (an accessary of A-500HS) under CO (Figure 2c; left). The oxygen indicator changes color from blue (>0.5% O_2) to pink (<0.1% O_2), and the degree of anaerobicity in the bag can be easily checked by eye. Crystallization was carried out at 20 °C. A number of thin crystals normally appeared within a few days (Figure 2b). These plate-like crystals are isomorphous with the biphyramidal ones, both belonging to a space group of tetragonal $P4_12_12$ with unit cell parameters of $a = b = 53.1$ Å, $c = 191.5$ Å. We note here that the thin crystals are characterized by a well-developed (010) (or ac) crystal face, while the bipyramidal ones by a four-fold screw axis (c-axis) passing through the apices of the two pyramids.

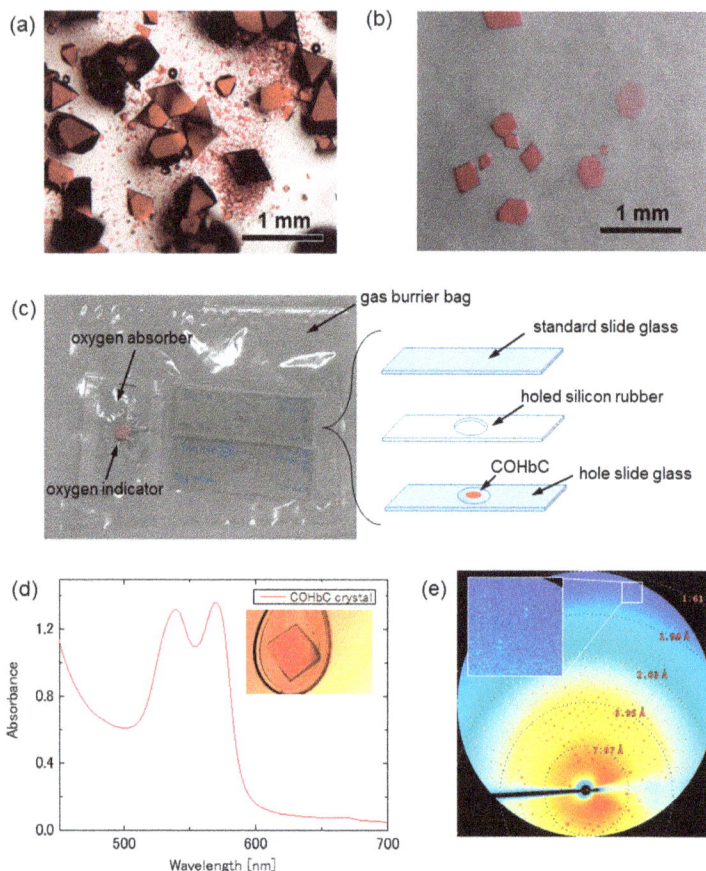

Figure 2. High quality, thin R-state COHbC crystals for X-ray crystallography (XRC) combined with CO photolysis and/or spectrophotometry. (**a**) COHbC crystals obtained by a conventional batch method; (**b**) Shape-controlled thin COHbC crystals; (**c**) Shape control approach using a spatially restricted crystallization; (**d**) Typical absorption spectrum of the shape-controlled thin COHbC crystal at 120 K. Inset is a photograph of the measured crystal; and, (**e**) An X-ray diffraction pattern of the same COHbC crystal as used for the spectral measurement. Diffraction spots are visible up to 1.7 Å.

The absorption spectrum of the shape-controlled thin COHbC crystal at 120 K, measured with unpolarized light incident on the (010) crystal face, is shown in Figure 2d, which includes a photograph of the measured crystal in the inset. Based on a millimolar extinction coefficient of 13.4 (per heme) at 540 nm for COHb and the calculated heme concentration in the crystal (i.e., about 50 mM), the thickness of the crystal can be roughly estimated to be 20 μm. An X-ray diffraction image of the same COHbC crystal as used for the spectral measurement is shown in Figure 2e. We confirmed that the diffraction spots are visible up to 1.7 Å.

2.1.2. Bezafibrate-Bound Horse COHb (BZF-COhHb) Crystal (R-State)

Another, more promising crystalline sample for the CO photolysis experiment is the crystal of CO-bound horse Hb (COhHb) in complex with the allosteric effector bezafibrate (BZF) (see chemical structure in Figure 3a). BZF was initially reported to reduce the oxygen affinity of Hb by preferentially binding and stabilizing the low-affinity T-state [25]. However, our previous study demonstrated that BZF can also bind to R-state COhHb to form a stable complex with slightly a lower oxygen affinity than free R-state [26]. An important observation was that a relatively thin, flat diamond-shaped BZF-COhHb crystal diffracts to 1.55 Å and gives a diffraction pattern with round, well defined diffraction spots, suggesting its potential usefulness for the photolysis experiment.

The very thin R-state BZF-COhHb crystals were made by a simple batch method using a gas-tight glass vial SVG-5 (Nichiden-Rika Glass) for gas-chromatography (Figure 3b). First, a mixed solution of 1.12 mL of 50% (w/v) polyethyleneglycol (PEG) 1000 and 0.88 mL of distilled water was placed in the vial and saturated with CO at 0 °C by CO bubbling on ice. Then, 4 mg of solid dithionite was added, followed by the addition of 2.0 mL of a filtered, CO-saturated 0.975% horse COHb solution in 0.1 M MES/NaOH buffer (pH 6.3) containing 16 mM BZF. The gas phase in the vial was quickly replaced by CO gas. Crystallization was carried out at 4 °C. A large number of very thin, flat diamond-shaped BZF-COhHb crystals appeared within a few days (Figure 3c).

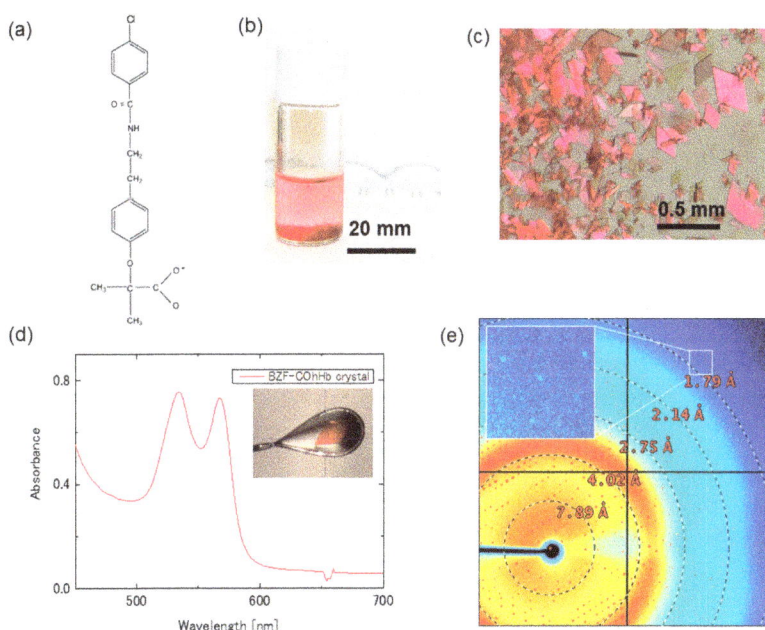

Figure 3. High quality, very thin R-state Bezafibrate-Bound Horse COHb (BZF-COhHb) crystals for XRC combined with CO photolysis and/or spectrophotometry. (**a**) The allosteric effector, bezafibrate (BZF); (**b**) Batch samples at 5 days after the crystallization setup; (**c**) Very thin BZF-COhHb crystals under polarized light; (**d**) Typical absorption spectrum of the BZF-COhHb crystal at 100 K. Inset shows a photograph of the measured crystal; and, (**e**) An X-ray diffraction pattern of the same COHbC crystal as used for the spectral measurement. Diffraction spots are visible at least to 1.7 Å.

Absorption spectrum at 100 K of the very thin BZF-COhHb crystal is shown in Figure 3d. Inset is a photograph of the measured crystal. The estimated thickness of the crystal is about 10 μm. A diffraction image of the same crystal as used for the spectral measurement is shown in Figure 3e. Diffraction spots are visible up to 1.7 Å (Figure 3e, Inset). This high resolution is very unusual and attractive, considering its thickness of only 10 μm (corresponding to the half thickness of the COHbC crystal as described above). The BZF-COhHb crystals are orthorhombic (space group C222$_1$) with unit cell parameters of a = 62.4 Å, b = 107.5 Å, c = 86.7 Å.

Adachi et al. [7] and Schotte et al. [8] have independently conducted X-ray analysis of the photoproduct of R-state COHb. These groups reported a 2.5 Å-resolution cryo-trapped photoproduct structure at 35 K, and time-resolved 2.0 Å-resolution electron density maps of the crystal at 15 °C, respectively. In both studies, the bipyramidal-shaped crystals of COHbA were used (with size of ~30 μm in the former study [7] and ~250 μm in the latter [8]). The yields of photoproducts at 35 K and 15 °C were about 50% and 15%, respectively. In both of the studies, the most prominent changes observed were the weakening of the electron density of the bound CO and the appearance of a new density of the photolyzed CO which is located near the heme. The changes in the atomic positions of the polypeptide moiety were very small, indicating no sign of allostric movements. The absence of protein conformational changes at 35 K is reasonably expected because large-scale protein fluctuations become frozen below ~180 K [27,28]. However, the absence of main chain motions at 15 °C could have resulted from the low photoproduct yield of 15%; a simple Monod-Wyman-Changeux (MWC) two-state model [29] suggests that 2 or 3 ligands (out of 4 per Hb molecule) must be photodissociated to trigger the R-to-T transition of tetrameric Hb. There is a possibility of detecting more conformational changes above 180 K by using a high-quality, thin COHb crystal. Our newly developed crystal forms, namely the shape-controlled thin COHbC crystal and the very thin BZF-COhHb crystal, are therefore a promising alternative to the well-known bipyramidal COHbA crystals for the observation of the initial allosteric movements in the Hb tetramer. Such attempts are now in progress in our laboratory.

2.2. Large-Volume (>20 mm^3) Hb Crystal Growth for Neutron Crystallgoraphy (NC)

2.2.1. DeoxyHbA Crystal (T-State)

Neutron crystallography (NC) is the prevailing technique for the accurate determination of the positions of hydrogen atoms in proteins, enabling the identification of the protonation states of amino acid residues and the nature of hydrogen bonds and salt-bridges [11,12]. While informative, this method requires large and well-ordered single crystals because available neutron sources are very weak. In most cases, crystals of larger than 0.5 mm^3 have been used to overcome the weak neutron flux of beamlines [30]. A further problem with NC is an anomalously large incoherent scattering from hydrogen (^{1}H) [31]. This scattering does not contribute to Bragg diffraction peaks, but rather produces a uniform background. On the other hand, the incoherent scattering cross section of deuterium (^{2}D) is approximately 40 times lower than that of ^{1}H, and thus it is necessary to grow crystals from, or soak the crystals in, D$_2$O solutions [31].

We accordingly modified the batch method of Perutz [32] to obtain large-sized, deuterated deoxyHbA crystals from D$_2$O solutions. Before crystallization, the stock solution of COHbA (in H$_2$O) was converted to the oxy-form by illumination for 30 min under a continuous stream of pure oxygen gas at 0 °C. The resulting oxyHbA sample was then diluted about tenfold with D$_2$O and concentrated by ultrafiltration at 5 °C, by using an Amicon ultrafiltration unit equipped with a YM-10 membrane (Millipore). The dilution and concentration were repeated until the free H$_2$O decreased down to less than 1%.

The concentrated oxyHbA solution was mixed with deuterated "solution C" of Perutz [32] to give a 8 mL solution of 1.0% (w/v) deoxyHbA, containing 4.30 M NH$_4$, 1.94 M SO$_4$, and 0.242 M PO$_4$ (pH 6.3; which was read straight from the pH meter without correction for the deuterium shift). Note that Perutz's solution C is a mixture of 4 M (NH$_4$)$_2$SO$_4$, 2 M (NH$_4$)H$_2$PO$_4$ and 2 M (NH$_4$)$_2$HPO$_4$ with a volume ratio of 0.8:0.05:0.15 (pH 6.5). We also note here that the solubility of oxyHbA is significantly higher than that of deoxyHbA under these solution conditions, and the oxyHbA solution is clear at this stage.

The deuterated oxyHbA sample was deoxygenated by repeated cycles of evacuation and equilibration with pure nitrogen gas at room temperature, then residual oxygen was scavenged by adding a small amount of solid sodium dithionite to give a final concentration of about 5 mM. The deoxygenated sample was placed into a gas-tight glass vial SVG-5 (Nichiden-Rika Glass) and quickly sealed with an oxygen absorber A-500HS (ISO, Yokohama, Japan) to maintain complete anaerobic conditions during crystallization at 20 °C. Crystals normally grew to their maximum size in about one month (Figure 4a). In our experience, the success rate of obtaining crystals of above 20 mm^3 in volume (with edge dimensions of ~3 mm) is about 50%.

Since neutron diffraction experiments require a long data collection time (from days to months), the deoxyHbA crystals must be kept wet and free from oxygen during this time period. Accordingly, a large crystal (4 × 3 × 3 mm^3) was mounted in a specially designed, anaerobic quartz capillary containing mother liquor (4.73 M NH$_4$, 2.13 M SO$_4$, and 0.267 M PO$_4$) in contact with the bottom of the crystal (Figure 4b). This capillary also contains ~100 mm^3 of iron powder of oxygen absorber A-500HS (ISO, Yokohama, Japan) and a small amount crushed oxygen indicator placed ~20 mm above the crystal (Figure 4b), to maintain the anaerobic conditions during the long neutron experiment. The capillary was tightly sealed with a rubber septum at the upper end (Figure 5a). This operation was carried out in a glove box filled with pure nitrogen. Finally, the upper end of the capillary was further sealed with epoxy resin (Figure 5a). Using this method, the deoxyHbA crystals can be kept wet and free from O$_2$ for more than a year.

The first neutron diffraction data on the deoxyHbA crystal to 2.1 Å resolution were collected at JRR-3M reactor in Japan Atomic Energy Agency (JAEA) using the BIX-3 diffractometer [33]. The neutron beam size was 5 mm in diameter and the wavelength was 2.9 Å. A step-scan data-collection method with an interval of 0.3° between frames was used. Two data sets of 588 and 621 still frames were taken with the different rotation axes. Exposure times were about 110 min per frame and the total time required was 120 days. The typical experimental set up and the neutron diffraction pattern of the deoxyHbA crystal are shown in Figure 5. The details of the neutron structure were described in elsewhere [33].

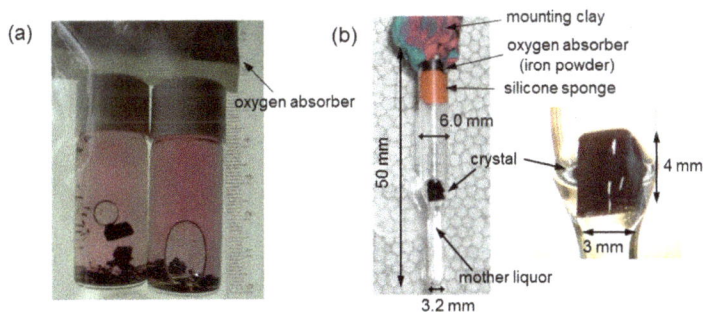

Figure 4. Large-volume, deuterated T-state deoxyHbA crystals for neutron crystallography (NC). (**a**) Batch samples at 55 days after the crystallization setup; (**b**) The sample for NC measurement. A deoxyHbA crystal of about 4 × 3 × 3 mm in size (~36 mm^3) was mounted in an anaerobic quartz capillary.

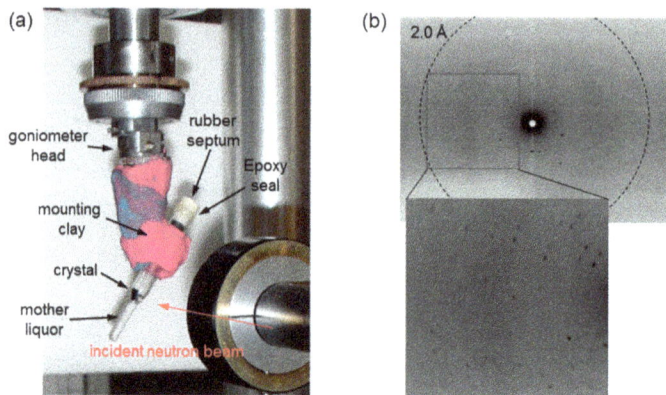

Figure 5. NC measurement for the T-state deoxyHbA crystal. (**a**) Experimental setup; (**b**) Typical neutron diffraction pattern of the deoxyHbA crystal. The dashed circle represents the resolution at 2.0 Å. The picture below is a close up view.

2.2.2. COHbA Crystal (R-State)

The oxygen affinity of Hb is lowered by the uptake of protons, but the mechanism of this alkaline Bohr effect is still not fully understood [34]. A comparison of the neutron structures of T-state deoxyHbA and R-state liganded Hb (e.g., oxyHbA or COHbA) provide direct information about the protonation changes in Hb associated with ligand binding, allowing identification of the Bohr protons. Thus, we further attempted the growth of large volume COHbA crystals for NC.

We combined the batch method of Perutz [32] and a seeding technique to obtain large-sized, deuterated COHbA crystals from D_2O solutions. First, deuteration of the stock solution of COHbA (in H_2O) was carried out according to the same procedure as described in Section 2.2.1 without converting to the oxy-form. The deuterated COHbA solution was mixed with deuterated 4 M phosphate buffer of Perutz [32] to give a 3 mL solution of 0.66% (*w*/*v*) COHbA in CO-saturated 2.10 M phosphate buffer containing 0.875 M NaH_2PO_4, 1.225 M K_2HPO_4, and 7.4 mM DL-homocysteine in D_2O. Note that Perutz's phosphate buffer is a mixture of 4 M NaH_2PO_4 and 4 M K_2HPO_4 with a volume ratio of 5:7 (pH 6.7). We also note here that the COHbA solution is almost clear at this stage.

Then, a small amount of the mother liquor of the above mentioned COHbC crystal sample (containing submicrometer-sized crystals and/or crystalline particles) was added to the COHbA solution as seeds, followed by the addition of 1 mL of paraffin oil and 100 µL of toluene. It is important to note that the addition of paraffin oil and toluene is effective for the growth of well-shaped crystals. The sample was placed in a gas-tight glass vial SVG-5 (Nichiden-Rika Glass) and quickly sealed with oxygen absorber A-500HS (ISO, Yokohama, Japan) under CO. Crystallization was carried out at 20 °C. A number of visible crystals appeared within several days at the oil-solution interface and normally grew to about 1 mm in size within one month (Figure 6a). Among them, well-formed single crystals were used as seeds for the next macroseeding step.

(a)   (b)

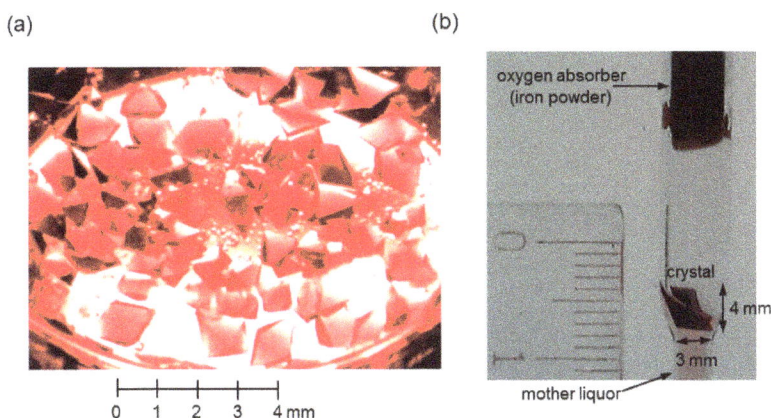

Figure 6. Large-volume, deuterated R-state COHbA crystals for NC. (**a**) The batch sample at 14 days after the first crystallization setup; (**b**) The COHbA crystal grown to about 4 × 3 × 2 mm in size (~24 mm³) at 368 days after the macroseeding setup. The crystal was mounted in a CO-filled anaerobic quartz capillary for NC.

Each selected COHbA crystal was soaked and washed several times in a drop of CO-saturated 2.00 M phosphate buffer containing 0.833 M NaH_2PO_4 and 1.167 M K_2HPO_4 in D_2O, where the COHbA crystals remain intact for a while but very slowly dissolve with time. The washed crystal was then transferred into a 6 mL solution of 0.66% (*w/v*) COHbA in CO-saturated 2.10 M deuterated phosphate buffer (as mentioned above) in a gas-tight glass vial SVG-30 (Nichiden-Rika Glass), followed by the gentle addition of 2 mL of paraffin oil and 200 μL of toluene. The vial was quickly sealed with oxygen absorber A-500HS (ISO, Yokohama, Japan) under CO, and crystal growth was carried out at 20 °C. Crystals grew to their maximum size in several months. Figure 6b shows an example of the grown COHbA crystal (4 × 3 × 2 mm³) mounted in a CO-filled anaerobic quartz capillary as used for the deoxyHbA crystals. Note that the NC measurement on this sample has not yet been completed.

Taken together, we have succeeded in obtaining large volume crystals of both deoxyHbA and COHbA for NC, by using approaches that are based on the classical batch method under optimized conditions. Macroseeding was also employed to enlarge the COHbA crystals. The obtained large-volume (>20 mm³) Hb crystals are one of the largest protein crystals ever grown for NC [30,31]. It is noteworthy that, in the case of Hb, the batch method is advantageous over the more commonly used crystallization methods, such as the vapor diffusion [35] and counter-diffusion methods [36], in terms of better anaerobicity and of higher capacity. In particular, completely anaerobic conditions are required to avoid oxidization of the ferrous heme to the ferric (or met) form when crystallizing either deoxyHbA or COHbA over a period of months. Our batch methods were designed to meet such a requirement.

2.3. Extremely Large-Volume (>100 mm³) Hb Crystal Growth for X-ray Fluorescence Holography (XFH)

X-ray fluorescence holography (XFH) is a novel imaging technique that utilizes a specific fluorescing metal as a wave source to monitor the interference field formed in a crystalline sample [37,38]. When compared to traditional diffraction-based XRC, a remarkable feature of XFH is that it can record both intensity and phase information, allowing model-free image reconstruction of the surrounding atoms through a simple Fourier-like transform. However, one negative aspect of this method is the very low signal-to-noise ratio (about 0.1%). While XFH has recently been successfully applied for local structural analysis of inorganic materials, its application to metalloprotein crystals remains limited, largely because protein crystals contain a much smaller proportion of metal atoms than inorganic crystals.

Hb is an attractive protein sample for XFH, as it contains one Fe(II)-heme group per subunit and crystals can be grown to very large sizes. However, since the size requirement for XFH appears to be much higher than that for NC, it is still necessary to grow the largest crystals possible. In addition, unlike NC, protein crystals must be kept cool throughout the XFH experiment to reduce X-ray radiation damage, requiring the presence of a cryoprotectant in the crystal growth conditions. Deuteration, or any other form of labeling, is not necessary for XFH.

With these requirements in mind, we optimized the method for COHbA crystallization (as described in Section 2.2.2) by adding glycerol (as a cryoprotectant) and using macroseeding for further crystal growth. First, the COHbA solution (in H_2O) was mixed with 4 M phosphate buffer of Perutz [32] to give a 2 mL solution of 2.0% (w/v) COHbA in CO-saturated 2.15 M phosphate buffer containing 0.896 M NaH_2PO_4, 1.254 M K_2HPO_4, 10% (v/v) glycerol, and 7.4 mM DL-homocysteine. Then, a small amount of the mother liquor of the COHbC crystal sample (as described in Section 2.1.1) was added to the COHbA solution as seeds, followed by the addition of 0.65 mL of paraffin oil and 0.1 mL of toluene. The COHbA sample was placed in a gas-tight glass vial SVG-5 (Nichiden-Rika Glass) and quickly sealed under CO. Crystallization was carried out at 20 °C. A number of visible crystals appeared within 24 h at the oil-solution interface.

After several days of crystallization, a qualified, medium-sized COHbA crystal (with the edge dimension of about 0.2 mm) was soaked and washed several in a drop of CO-saturated 2.00 M phosphate buffer containing 0.833 M NaH_2PO_4, 1.167 M K_2HPO_4, and 10% (v/v) glycerol. The washed crystal was then transfer into a 2 mL solution of 2.0% (w/v) COHbA in CO-saturated 2.15 M phosphate buffer (as mentioned above) in a gas-tight glass vial SVG-5 (Nichiden-Rika Glass), followed by the gentle addition of 0.65 mL of paraffin oil and 0.1 mL of toluene. The vial was quickly sealed under CO, and crystal growth was carried out at 20 °C. Crystals grew to about 5 mm in size within one month (Figure 7a,b).

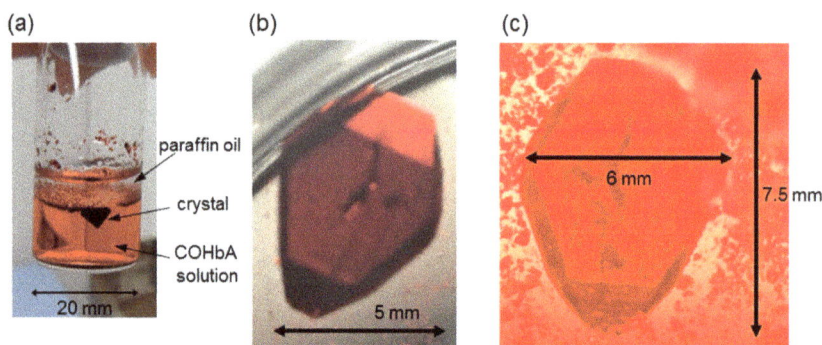

Figure 7. An extremely large-volume R-state COHbA crystal for XFH. (**a**) Example of the batch method using paraffin oil. Crystals appeared at the oil-solution interface (**b**) The COHbA crystal at 34 days after the first macroseeding setup. The crystal was washed to remove tiny crystals attached on the surface of the large crystal; (**c**) The same COHbA crystal grown to about 7.5 × 6 × 3 mm in size (~130 mm^3) at 193 days after the second macroseeding setup. This crystal was used for the measurement of XFH (see Figure 8a).

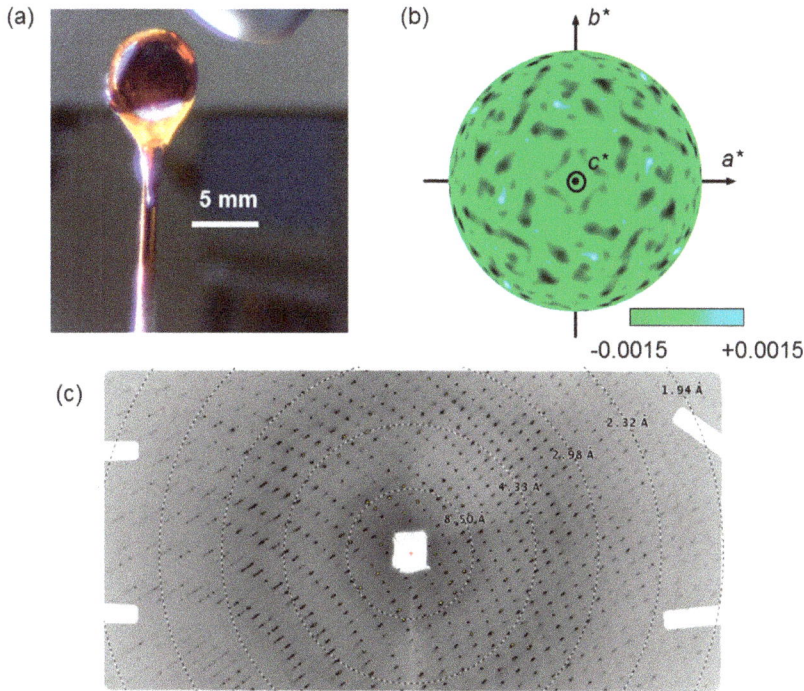

Figure 8. X-ray fluorescent holography (XFH) measurement for the COHbA crystal. (**a**) Crystal setup for data collection. Our largest COHbA crystal of about $7.5 \times 6 \times 3$ mm in size (~130 mm^3) was mounted on a customized loop made by a polyimide film (0.125 mm thickness) with a copper wire (1.2 mm diameter), and cooled to 100 K by nitrogen gas flow; (**b**) Typical hologram pattern of the COHbA crystal (excited by 7.25 keV X-ray photons), represented as a stereographic projection of an interference pattern in the reciprocal space (a^*, b^*, c^*) after applying a Gaussian low path spatial filter [13]; (**c**) An X-ray diffraction pattern of the COHbA crystal used for the XFH measurement. The measurement was carried out using an imaging plate of 2048×4020 pixels. The wavelength of X-ray was 1.127 Å.

To further grow this crystal, the second macroseeding was carried out. The crystal was soaked and washed several times in a drop of CO-saturated 2.00 M phosphate buffer containing 0.833 M NaH$_2$PO$_4$, 1.167 M K$_2$HPO$_4$, and 10% (v/v) glycerol (Figure 7b), and then transfer into a 10 mL solution of 1.5% (w/v) COHbA in CO-saturated 2.15 M phosphate buffer (as mentioned above) in a gas-tight glass vial SVG-30 (Nichiden-Rika Glass), followed by the gentle addition of 3.25 mL of paraffin oil and 0.5 mL of toluene. The vial was quickly sealed under CO, and crystal growth was carried out at 20 °C. Crystals grew to over 100 mm^3 in volume within several months (Figure 7c). To our knowledge, this crystal size is far larger than any previously reported sizes of Hb crystals.

For the XFH measurement, the large COHbA crystal was briefly soaked in CO-saturated 2.30 M phosphate buffer containing 0.958 M NaH$_2$PO$_4$, 1.342 M K$_2$HPO$_4$, and 15% (v/v) glycerol, and then flash cooled to 100 K to reduce X-ray radiation damage (Figure 8a). The first XFH data-set using a COHbA crystal was collected at the beamline BL6C of Photon Factory [13]. A step-scan data collection was performed with an interval of $0.25°$ and $1.0°$ in pitch and yaw, respectively. The data were collected using eight X-ray energies in the 7.25–10.75 keV region in a step of 0.5 keV. A typical hologram pattern of the COHbA crystal is shown in Figure 8b. We checked the crystal quality by measuring the Bragg diffraction pattern for X-rays incident on the crystal (Figure 8c). More details of the XFH measurement are described in elsewhere [13]. Note that the crystals grown at the oil-solution interface tend to have shapes with well-developed crystal faces (Figures 7b,c and 8a). This shape is well suited to minimize the "crystal shape effect" on the fluorescent X-ray intensity during crystal rotation necessary for the XFH measurement.

3. Materials and Methods

3.1. Preparation of HbA, HbC, and Horse Hb

HbA was prepared and purified in the CO form, as reported previously [6]. HbC was prepared and purified in the CO form by a method described previously [10]. Horse Hb was prepared in the CO form as reported previously [26].

3.2. Microspectrophotometry and Diffraction Image Collection for the COHbC Crystal

Microspectrophotometry of the R-state COHbC crystal was performed by a similar system, as reported previously [39], except using a different microscope (Eclipse TE2000-U, Nikon, Tokyo, Japan), spectrometer (double monochromator, Horiba Jobin Yvon, Kyoto, Japan), photomultiplier (model C3830, Hamamatsu Photonics, Hamamatsu, Japan), light chopper (5584A, NF, Yokohama, Japan) and digital lock-in amplifier (LI5640, NF, Yokohama, Japan). The crystal was cooled to 120 K by cold nitrogen gas (open-flow cryostat, Rigaku, Tokyo, Japan). The mother liquor containing 15% (v/v) glycerol was used for the COHbC crystal as cryoprotectant, in which the crystal was rinsed briefly before flash-freezing in liquid nitrogen.

The diffraction images were collected at the beamline NW14A at PF-AR, High Energy Accelerator Research Organization (KEK) using CCD detector (marccd, rayonix, Quebec, QC, Canada) and mardtb stage [40]. The crystal was cooled to 140 K by cold nitrogen gas (cryostream 600, Oxford cryosystems, Oxford, UK). X-ray wavelength, detector distance, oscillation range, and 2theta angle are 0.827 Å, 250 mm, 1° and 12°, respectively.

3.3. Microspectrophotometry and Diffraction Image Collection for the BZF-COhHb Crystal

On-line microspectrophotometry of the R-state BZF-COhHb crystal was performed at the beamline X26C at NSLS, Brookhaven National Laboratory using UV-VIS light source (L10290, Hamamatsu Photonics, Hamamatsu, Japan) and spectrometer (QE65000, Ocean Optics, Largo, FL, USA) [41]. The crystal was cooled to 100 K by cold nitrogen gas (cryostream 600, Oxford cryosystems, Oxford, UK). The mother liquor containing 20% (v/v) glycerol was used for the BZF-COhHb crystal as cryoprotectant, in which the crystal was rinsed briefly before flash-freezing in liquid nitrogen.

The diffraction image was collected at the beamline X26C using CCD detector (315r, Quantum detector, Oxford, UK). X-ray wavelength, detector distance, and oscillation range are 1.0 Å, 250 mm and 1°, respectively. The crystal was cooled to 100 K.

4. Conclusions

This work illustrates the various shapes and sizes of Hb crystals that can be grown by carefully controlling and optimizing the crystallization conditions by using the knowledge accumulated in our laboratory. Methods are described for obtaining reproducibly high-quality, thin crystals, and growing crystals with dimensions over 5 mm. The quality of these crystals and their applicability to novel crystallographic techniques were also examined. These Hb crystals are suitable for X-ray crystallography (XRC) combined with photoexcitation and/or spectrophotometry, neutron crystallography (NC), and X-ray fluorescent holography (XFH). Although beyond the scope of this article, it is important to note that one of the common challenges now faced is growing very small crystals for use with X-ray free electron laser (XFEL) serial crystallography experiments. We expect that the methods presented here will provide a basis for new and challenging experiments, and will encourage further progress of protein crystallographic work that poses very strict requirements on crystal size and quality.

Acknowledgments: We thank K. Hayashi, N. Happo and Y. C. Sasaki for help with the XFH experiment, A. M. Orville, B. Andi, T. Senda, M. Senda, S. Adachi, S. Nozawa and T. Sato for help with the XRC and spectroscopic experiments, and T. Chatake and Y. Morimoto for help with the NC experiment. We also thank J. Tame for helpful comments and suggestions to improve the paper. This work was supported by the MEXT/JSPS Grants-in-Aid for JSPS Fellows 11J10227 (to A.S.-T.), JSPS KAKENHI Grant Numbers 26105005 (to N.S.), 26840028 (to A.S.-T.), JP15H01646 (to N.S.), JP16K07326 (to N.S.), JP17H06372 (to A.S.-T.), and Jichi Medical University Young Investigator Award (to A.S.-T.). The XRD experiments were performed at the beamlines X26C of NSLS and NW14A of Photon Factory Advanced Ring (PF-AR) with the approval of KEK (Proposal No. 2009G683 and 2011G607). The XFH experiments were performed at the beamline BL6C of Photon Factory with the approval of KEK (Proposal No. 2013G605, 2013G653, and 2015G589).

Author Contributions: Naoya Shibayama designed the research, carried out Hb preparation and most of crystallization, performed experiments, and wrote the manuscript. Ayana Sato-Tomita carried out crystallization, performed experiments and generated data and figures. Both authors contributed to editing and reviewing the manuscript.

Conflicts of Interest: The authors declare no conflict of interest.

References

1. Kendrew, J.C.; Bodo, G.; Dintzis, H.M.; Parrish, R.G.; Wyckoff, H.; Phillips, D.C. A three-dimensional model of the myoglobin molecule obtained by X-ray analysis. *Nature* **1958**, *181*, 662–666. [CrossRef] [PubMed]

2. Pertz, M.F.; Rossmann, M.G.; Cullis, A.F.; Muirhead, H.; Will, G.; North, A.C. Structure of haemoglobin: A three-dimensional Fourier synthesis at 5.5-A. resolution, obtained by X-ray analysis. *Nature* **1960**, *185*, 416–422. [CrossRef]

3. Kendrew, J.C.; Dickerson, R.E.; Strandberg, B.E.; Hart, R.G.; Davies, D.R.; Phillips, D.C.; Shore, V.C. Structure of myoglobin: A three-dimensional Fourier synthesis at 2 Å resolution. *Nature* **1960**, *185*, 422–427. [CrossRef] [PubMed]

4. RCSB Protein Data Bank. Available online: https://www.rcsb.org (accessed on 2 September 2017).

5. Shibayama, N.; Morimoto, H.; Kitagawa, T. Properties of chemically modified Ni(II)-Fe(II) hybrid hemoglobins: Ni(II) protoporphyrin IX as a model for a permanent deoxy-heme. *J. Mol. Biol.* **1986**, *192*, 331–336. [CrossRef]

6. Shibayama, N.; Imai, K.; Hirata, H.; Hiraiwa, H.; Morimoto, H.; Saigo, S. Oxygen equilibrium properties of highly purified human adult hemoglobin cross-linked between $82\beta1$ and $82\beta2$ lysyl residues by bis(3,5-dibromosalicyl)fumarate. *Biochemistry* **1991**, *30*, 8158–8165. [CrossRef] [PubMed]

7. Adachi, S.; Park, S.-Y.; Tame, J.R.H.; Shiro, Y.; Shibayama, N. Direct observation of photolysis-induced tertiary structural changes in hemoglobin. *Proc. Natl. Acad. Sci. USA* **2003**, *100*, 517–520. [CrossRef] [PubMed]

8. Schotte, F.; Cho, H.S.; Soman, J.; Wulff, M.; Olson, J.S.; Anfinrud, P.A. Real-time tacking of CO migration and binding in the α and β subunits of human hemoglobin via 150-ps time-resolved Laue crystallography. *Chem. Phys.* **2013**, *422*, 98–106. [CrossRef] [PubMed]

9. Mozzarelli, A.; Rivetti, C.; Rossi, G.L.; Henry, E.R.; Eaton, W.A. Crystals of haemoglobin with the T quaternary structure bind oxygen noncooperatively with no Bohr effect. *Nature* **1991**, *351*, 416–419. [CrossRef] [PubMed]

10. Shibayama, N.; Sugiyama, K.; Park, S.Y. Structures and oxygen affinities of crystalline human hemoglobin C ($\beta6$ Glu→Lys) in the R and R2 quaternary structures. *J. Biol. Chem.* **2011**, *286*, 33661–33668. [CrossRef] [PubMed]

11. Blakeley, M.P.; Langan, P.; Niimura, N.; Podjarny, A. Neutron crystallography: Opportunities, challenges, and limitations. *Curr. Opin. Struct. Biol.* **2008**, *18*, 593–600. [CrossRef] [PubMed]

12. Niimura, N.; Bau, R. Neutron protein crystallography: Beyond the folding structure of biological macromolecules. *Acta Crystallogr.* **2008**, *64*, 12–22. [CrossRef] [PubMed]

13. Sato-Tomita, A.; Shibayama, S.; Happo, N.; Kimura, K.; Okabe, T.; Matsushita, T.; Park, S.-Y.; Sasaki, Y.C.; Hayashi, K. Development of an X-ray fluorescence holographic measurement system for protein crystals. *Rev. Sci. Instrum.* **2016**, *87*. [CrossRef] [PubMed]

14. Dickerson, R.E.; Geis, I. *Hemoglobin: Structure, Function, Evolution, and Pathology*, 1st ed.; The Benjamin/ Cummings Publishing Company: Menlo Park, CA, USA, 1983; pp. 76–116, ISBN 0-805-32411-9.

15. Baldwin, J.; Chothia, C. Haemoglobin: The structural changes related to ligand binding and its allosteric mechanism. *J. Mol. Biol.* **1979**, *129*, 175–220. [CrossRef]

16. Ren, Z. Reaction trajectory revealed by a joint analysis of protein data bank. *PLoS ONE* **2013**, *8*, e77141. [CrossRef] [PubMed]

17. Schlichting, I.; Berendzen, J.; Phillips, G.N., Jr.; Sweet, R.M. Crystal structure of photolysed carbonmonoxy-myoglobin. *Nature* **1994**, *371*, 808–812. [CrossRef] [PubMed]

18. Teng, T.Y.; Srajer, V.; Moffat, K. Photolysis-induced structural changes in single crystals of carbonmonoxy myoglobin at 40 K. *Nat. Struct. Biol.* **1994**, *1*, 701–705. [CrossRef] [PubMed]

19. Ostermann, A.; Waschipky, R.; Parak, F.G.; Nienhaus, G.U. Ligand binding and conformational motions in myoglobin. *Nature* **2000**, *404*, 205–208. [CrossRef] [PubMed]

20. Tomita, A.; Sato, T.; Ichiyanagi, K.; Nozawa, S.; Ichikawa, H.; Chollet, M.; Kawai, F.; Park, S.Y.; Tsuduki, T.; Yamato, T.; et al. Visualizing breathing motion of internal cavities in concert with ligand migration in myoglobin. *Proc. Natl. Acad. Sci. USA* **2009**, *106*, 2612–2616. [CrossRef] [PubMed]

21. Sawicki, C.A.; Gibson, Q.H. Dependence of the quantum efficiency for photolysis of carboxyhemoglobin on the degree of ligation. *J. Biol. Chem.* **1979**, *254*, 4058–4062. [PubMed]

22. Hirsch, R.E.; Juszczak, L.J.; Fataliev, N.A.; Friedman, J.M.; Nagel, R.L. Solution-active structural alterations in liganded hemoglobins C (β6 Glu→Lys) and S (β6 Glu→Val). *J. Biol. Chem.* **1999**, *274*, 13777–13782. [CrossRef] [PubMed]

23. Diggs, L.W.; Kraus, A.P.; Morrison, D.B.; Rudnicki, R.P. Intraerythrocytic crystals in a white patient with hemoglobin C in the absence of other types of hemoglobin. *Blood* **1954**, *9*, 1172–1184. [PubMed]

24. Hirsch, R.E.; Raventos-Suarez, C.; Olson, J.A.; Nagel, R.L. Ligand state of intraerythrocytic circulating HbC crystals in homozygote CC patients. *Blood* **1985**, *66*, 775–777. [PubMed]

25. Perutz, M.F.; Fermi, G.; Abraham, D.J.; Poyart, C.; Bursaux, E. Hemoglobin as a receptor of drugs and peptides: X-ray studies of the stereochemistry of binding. *J. Am. Chem. Soc.* **1986**, *108*, 1064–1078. [CrossRef]

26. Shibayama, N.; Miura, S.; Tame, J.R.H.; Yonetani, T.; Park, S.-Y. Crystal structure of horse carbonmonoxyhemoglobin-bezafibrate complex at 1.55-Å resolution. A novel allosteric binding site in R-state hemoglobin. *J. Biol. Chem.* **2002**, *277*, 38791–38796. [CrossRef] [PubMed]

27. Parak, F.; Knapp, E.W.; Kucheida, D. Protein dynamics: Mössbauer spectroscopy on deoxymyoglobin crystals. *J. Mol. Biol.* **1982**, *161*, 177–194. [CrossRef]

28. Doster, W.; Cusack, S.; Petry, W. Dynamical transition of myoglobin revealed by inelastic neutron scattering. *Nature* **1989**, *337*, 754–756. [CrossRef] [PubMed]

29. Monod, J.; Wyman, J.; Changeux, J.-P. On the nature of allosteric transitions: A plausible model. *J. Mol. Biol.* **1965**, *12*, 88–118. [CrossRef]

30. Ng, J.D.; Baird, J.K.; Coates, L.; García-Ruiz, J.M.; Hodge, T.A.; Huang, S. Large-volume protein crystal growth for neutron macromolecular crystallography. *Acta Crystallogr. F Struct. Biol. Commun.* **2015**, *71*, 358–370. [CrossRef] [PubMed]

31. Blakeley, M.P.; Hasnain, S.S.; Antonyuk, S.V. Sub-atomic resolution X-ray crystallography and neutron crystallography: promise, challenges and potential. *IUCrJ* **2015**, *2*, 464–474. [CrossRef] [PubMed]

32. Pertz, M.F. Preparation of hemoglobin crystals. *J. Cryst. Growth* **1968**, *2*, 54–56. [CrossRef]

33. Chatake, T.; Shibayama, N.; Park, S.-Y.; Kurihara, K.; Tamada, T.; Tanaka, I.; Niimura, N.; Kuroki, R.; Morimoto, Y. Protonation states of buried histidine residues in human deoxyhemoglobin revealed by neutron crystallography. *J. Am. Chem. Soc.* **2007**, *129*, 14840–14841. [CrossRef] [PubMed]

34. Imai, K.; Yonetani, T. PH dependence of the Adair constants of human hemoglobin. Nonuniform contribution of successive oxygen bindings to the alkaline Bohr effect. *J. Biol. Chem.* **1975**, *250*, 2227–2231. [PubMed]

35. Benvenuti, M.; Mangani, S. Crystallization of soluble proteins in vapor diffusion for X-ray crystallography. *Nat. Protoc.* **2007**, *2*, 1633–1651. [CrossRef] [PubMed]

36. García-Ruiz, J.M. Counter diffusion methods for macromolecular crystallization. *Methods Enzymol.* **2003**, *368*, 130–154. [PubMed]

37. Tegze, M.; Faigel, G. X-ray holography with atomic resolution. *Nature* **1996**, *380*, 49–51. [CrossRef]

38. Gog, T.; Len, P.M.; Materlik, G.; Bahr, D.; Fadley, C.S.; Sanchez-Hanke, C. Multiple-energy X-ray holography: Atomic images of hematite (Fe_2O_3). *Phys. Rev. Lett.* **1996**, *76*, 3132–3135. [CrossRef] [PubMed]

39. Sakai, K.; Matsui, Y.; Kouyama, T.; Shiro, Y.; Adachi, S. Optical monitoring of freeze-trapped reaction intermediates in protein crystals: a microspectro-photometer for cryogenic protein crystallography. *J. Appl. Cryst.* **2002**, *35*, 270–273. [CrossRef]

40. Nozawa, S.; Adachi, S.; Takahashi, J.; Tazaki, R.; Guérin, L.; Daimon, M.; Tomita, A.; Sato, T.; Chollet, M.; Collet, E.; et al. Developing 100 ps-resolved X-ray structural analysis capabilities on beamline NW14A at the Photon Factory Advanced Ring. *J. Synchrotron Radiat.* **2007**, *14*, 313–319. [CrossRef] [PubMed]

41. Orville, A.M.; Buono, R.; Cowan, M.; Heroux, A.; Shea-McCarthy, G.; Schneider, D.K.; Skinner, J.M.; Skinner, M.J.; Stoner-Ma, D.; Sweet, R.M. Correlated single-crystal electronic absorption spectroscopy and X-ray crystallography at NSLS beamline X26-C. *J. Synchrotron Radiat.* **2011**, *18*, 358–366. [CrossRef] [PubMed]

crystals

MDPI

Article

Phenomenological Consideration of Protein Crystal Nucleation; the Physics and Biochemistry behind the Phenomenon

Christo N. Nanev

Rostislaw Kaischew Institute of Physical Chemistry, Bulgarian Academy of Sciences, 1113 Sofia, Bulgaria; nanev@ipc.bas.bg; Tel.: +359-2-856-6458; Fax: +359-2-971-2688

Academic Editor: Jolanta Prywer
Received: 23 May 2017; Accepted: 21 June 2017; Published: 27 June 2017

Abstract: Physical and biochemical aspects of protein crystal nucleation can be distinguished in an appropriately designed experimental setting. From a physical perspective, the diminishing number of nucleation-active particles (and/or centers), and the appearance of nucleation exclusion zones, are two factors that act simultaneously and retard the initially fast heterogeneous nucleation, thus leading to a logistic time dependence of nuclei number density. Experimental data for protein crystal (and small-molecule droplet) nucleation are interpreted on this basis. Homogeneous nucleation considered from the same physical perspective reveals a difference—the nucleation exclusion zones lose significance as a nucleation decelerating factor when their overlapping starts. From that point on, a drop of overall system supersaturation becomes the sole decelerating factor. Despite the different scenarios of both heterogeneous and homogeneous nucleation, S-shaped time dependences of nuclei number densities are practically indistinguishable due to the exponential functions involved. The biochemically conditioned constraints imposed on the protein crystal nucleation are elucidated as well. They arise because of the highly inhomogeneous (patchy) protein molecule surface, which makes bond selection a requisite for protein crystal nucleation (and growth). Relatively simple experiments confirm this assumption.

Keywords: protein crystallization; classical and two-step nucleation mechanisms; physical and biochemical aspects of protein crystal nucleation; S-shaped nucleation kinetics

1. Introduction

Biomolecule structures are essential when it comes to understanding the mechanisms of life and human genomes, and developing novel protein-based pharmaceuticals. The most powerful method for structure-function studies of biomolecules is X-ray diffraction (with complementary neutron diffraction) and Nuclear Magnetic Resonance, considered as an ancillary tool only. Both X-ray and neutron diffraction require well-diffracting crystals [1]. Growing such crystals of newly-expressed proteins is, however, the major obstacle in protein structure determination. There is no recipe for their growth. It is usually the trial-and-error approach that is applied. Despite the numerous state-of-the-art crystallization tools employed (such as robots, automation and miniaturization of crystallization trials, Dynamic Light Scattering, crystallization screening kits, etc.), researchers' creativeness and acumen remain indispensable.

Protein crystal nucleation is a prerequisite for the crystal growth of newly-expressed proteins. However, there is no theory that could help predict adequately crystallization conditions. Quite often, the classical nucleation theory (CNT) is employed to give a (physical) rendition of protein crystal nucleation process. While providing a logical explanation of the fluctuation-based mechanism and the origin of nucleation barrier, CNT fails to predict correctly nucleation rates. In some cases, the deviations are of many orders of magnitude, e.g., [2]. In this work, applying microfluidics technologies, localized DC electric field, and gel crystallization, the authors studied the spatial and temporal location of the nucleation event. They used a confinement effect coupled to an external localized DC electric field to evoke a desired nucleation and growth of lysozyme crystals, in 20 mg/mL lysozyme, 0.7 M NaCl in agarose gel 1%.

A reason for the inadequacy between some experiments and the CNT could be the uncertainty in determining the energy of the interface arising between the new phase and the mother phase—interface energy variation of only 10% can alter the nucleation rate substantially because it depends exponentially on the nucleation energy barrier, which in turn is determined by the interface free energy in power three. The issue with CNT lies in the assumption that an emerging nucleus already has the order and density of the bulk crystal. The interface is described as a sharp surface with a specific (per unit area) free energy, usually not available from direct measurements. However, Wölk et al. [3] have shown that in cases for which CNT was devised originally, such as homogeneous nucleation of water droplets, a simple empirical modification to the CNT-nucleation rate (expressed by Becker–Doering formula) yields a robust function for predicting water nucleation rates over broad ranges of temperature and supersaturation.

The so-called two-step nucleation mechanism (TSNM) denies the simultaneous densification and ordering during a single nucleation event. While preserving the CNT basic concept for a fluctuation-based nucleation mechanism, TSNM assumes nucleation initiation via an intermediate condensed liquid appearing in the bulk solution. Being only densified, the intermediate phase preserves some similarity to the mother phase. Therefore, the phase-transition energy barrier is lowered bellow the one needed for direct transition mother-phase-to-crystal occurring via the CNT mechanism. The second step in TSNM is the formation of crystal nuclei inside the highly-concentrated regions. Thus, TSNM resembles the Ostwald's rule of stages, which stipulates that a thermodynamically less-stable phase appears first, then a polymorphic transition toward a stable phase occurs. Ten Wolde and Frenkel [4] have predicted theoretically the existence of amorphous precursors that have been further confirmed experimentally by Vivares et al. [5], Sauter et al. [6], and Schubert et al. [7]. Sleutel and Van Driessche [8] have observed a non-classical nucleation for the 3D liquid-to-crystal transition of glucose isomerase—local increase in density and crystallinity do not occur simultaneously, but rather sequentially. They have demonstrated that at high concentrations (~100 mg/mL), glucose isomerase can form mesoscopic liquid-like aggregates (the molecules in them retain enough mobility), which are potential precursors of crystalline clusters. These aggregates are stable with respect to the parent liquid, and metastable compared with the crystalline phase. In contrast, glucose isomerase 2D crystal nucleation proceeds classically [9] and they proved the existence of a critical crystal size. They also observed that the interior of all clusters is in the crystalline state and the cluster dynamics are determined by single molecular attachment and detachment events. Whitelam presents a molecular model designed to study crystallization in the presence and absence of amorphous intermediates [10]. Based on computer simulation, he suggests tuning the relative strengths of the specific and nonspecific interactions. Thus, the relative efficiencies of the various pathways leading towards the final crystalline state have been studied. Most recently, direct transition electron microscopic observations of Yamazaki et al. [11] have suggested a significant departure from the initial TSNM assumption. The authors have never observed formation of crystalline phases inside amorphous solid particles consisting of lysozyme molecules, which are like those previously assumed to consist of a dense liquid.

Although governed by physical laws established previously for small molecule crystallization, protein crystal nucleation is an extremely complex process. The complexity arises from the subtle interplay between process physics and biochemistry. It is the large size of the protein molecules and their highly inhomogeneous and patchy surface [12] that make the molecular-kinetic mechanism of protein crystal nucleation so specific. Protein crystal nucleation rate is reduced by a biochemical constraint associated with the strict selection of crystalline bonds. Based on experiments, this paper differentiates physical from biochemical protein crystal nucleation aspects.

2. Results and Discussion

2.1. Experimental Results

Any attempt to formulate accurate predictions by amending and overcoming CNT limitations should rest upon interpretation of some basic experimental observations. For instance, experimental data show that nuclei number density (n) of a new phase (crystals, droplets) depends simultaneously on both time (t) and supersaturation ($\Delta\mu$), i.e., $n = n(t, \Delta\mu)$. S-shaped dependences of n vs. t at constant supersaturation have been known to cause electrochemical new-phase nucleation for a long time e.g., [13,14]. But they remained unelucidated [15] until recently, when it was shown that they obey logistic functional dependence [16]. The same function also governs insulin crystal nucleation—large amounts of data for which can be found in [17]. Using custom-made quasi-two-dimensional all-glass cells with intentionally introduced air bubbles, n vs. t dependences were measured in this study simultaneously at four typical places: in solution bulk, at the glass support, at the air/solution interface, and at the three-phase boundary solution/glass/air. Stationary nucleation rates were determined from the linear parts of the corresponding plots, and energy barriers for nucleus formation and nucleus sizes were estimated. By simply focusing the microscope on the upper glass plate of the cell, heterogeneous on-glass crystal nucleation is differentiated from the one in the bulk solution. It is also argued that the latter proceeds heterogeneously, on some (unknown) foreign particles of biological origin. Seven different supersaturations have been studied with BioChemika-insulin, showing that crystal nucleation in bulk solution prevails greatly [17].

Using digitalized original experimental data from [14,18], logistic dependences (with very high goodness of fit, R^2) are presented in Figures 1 and 2. Such time dependence has also been established for bovine β-lactoglobulin crystal nucleation which proceeds by a TSNM [6]. Good logistic fits of insulin crystal nucleation data for seven different supersaturations are shown in Figure 3, where appropriate (supersaturation dependent) parameters are used. The relations, showing the degrees to which saturated crystal-nuclei number densities (n_s) are neared, (n/n_s), are plotted vs. t/t_p (using Equation (2); here t_p is the time for reaching n_s; $t_p = 2t_c$, and t_c is the time when the half of n_s is reached (namely the mid-point of the corresponding sinusoid). Plots in Figure 4a are for bulk insulin crystal nucleation, and in Figure 4b—for on-glass crystal nucleation. This issue will be considered below.

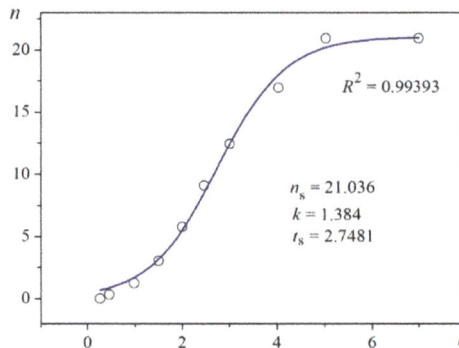

Figure 1. Logistic plot for mercury droplets, n (cm^{-2}) vs. t (msec); Figure 8, 84 mV in [14].

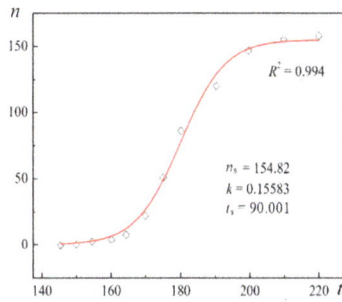

Figure 2. Logistic plot for lysozyme crystal nucleation on nanoporous gold; n (cm^{-3}) vs. t (min); Figure 4a in [18].

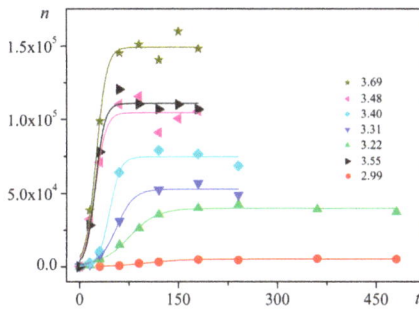

Figure 3. Experimental data for insulin crystal nucleation in bulk solution, n vs. t at series of dimensionless supersaturations, $\ln(c/c_e)$, where c is the actual insulin concentration, and c_e is the equilibrium concentration. The corresponding dimensionless supersaturations are given on the right-hand side. For the color references, refer to the web version of this article.

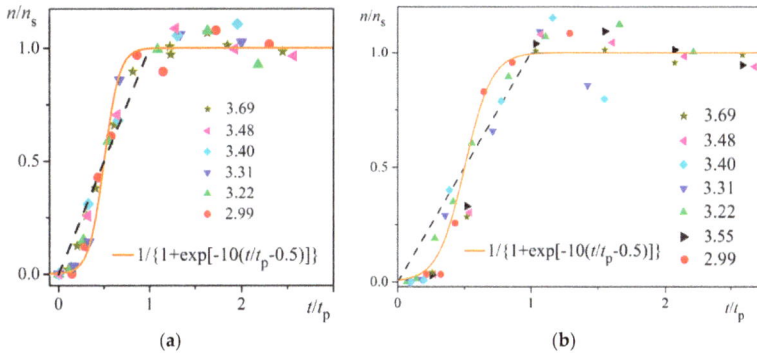

(a)

(b)

Figure 4. Logistic functional plots of n/n_s vs. t/t_p for insulin crystal nucleation. All experimental data for the dimensionless supersaturations studied (numbers on right-hand side) fall on the orange logistic curves (for the color references, refer to the web version of this article). The dashed straight lines with coordinates (00), (11) are a guide for the eye only; it is seen that the experimental points less than $t/t_p = 0.5$ are situated below the dashed straight line, while the points for greater t/t_p values (up to 1) lie above this line. Experimental data are plotted for: (a) bulk nucleation; (b) on-glass nucleation.

In conclusion, being common for both small inorganic molecules and large bio-molecules, the S-shaped dependences of nuclei number densities on time underline a common physical aspect of the nucleation processes.

2.2. Logistic Time Dependence of Protein and Small Molecule Crystal Nucleation

The fluctuation-based concept of CNT supposes that a new distribution of larger clusters starts replacing the equilibrium one immediately after establishing supersaturation in the system. Since the larger the cluster is, the longer it takes for it to emerge, the critically sized cluster should appear the latest. Importantly, the nucleus is the cluster of maximum energy and minimum concentration. Therefore, many subcritical clusters, smaller than the critical nucleus by a single molecule only, are formed in the meantime. However, the rearrangement of cluster size distribution does not end with the emergence of the very first critical cluster. Accommodation of the supersaturated system state continues gradually, leading to an enhanced supply of nuclei. Thus, the nucleation rate increases throughout the initial non-steady-state nucleation period.

As per definition, the momentary nucleation rate (dn/dt) during the initial non-steady-state nucleation period, i.e., the rate at any point of the n vs. t graph, is given by the number (n) of nuclei formed in a unit volume (1 cm^3) divided by the (infinitesimally short) nucleation time (t). Denoting the frequency of molecule attachments leading to formation of nuclei by k (s^{-1}), gives $dn/dt = kn$. Here, the attachment frequency k is defined as the frequency of molecule attachments to clusters which are smaller than the critical nucleus by a single molecule, *minus* the frequency of molecule detachments.

The attachment frequency k depends on supersaturation, which, however, remains constant during the whole nucleation process. The reason is that the extremely small nucleus volume (typically about 10^{-19} cm^3) and nucleation per se does not change the overall supersaturation—even during the most intensive nucleation (e.g., n approaching 10^6 cm^{-3}). Thus, beginning with a single nucleus, the nucleation process advances in an exponential manner with time. Nonetheless, no unlimited nuclei augmentation is physically feasible. Experimental results show that after a rapid initial increase, the nucleation process gradually decelerates to an almost constant nucleation rate up to reaching saturated nuclei number densities (n_s) in the plateau regions of the n vs. t dependences (Figure 3). Nucleation rate changes have been attributed [16] to two retardation factors acting simultaneously for heterogeneous nucleation (different for homogeneous nucleation).

A basic assumption of CNT is the supposition of continuous cluster size changes, which is a good approximation to reality only for large critical clusters. The consideration presented here does not suffer from such a limitation—irrespective of the mechanism involved, either CNT or TSNM, it is capable of accounting for discrete cluster size changes as well.

2.2.1. Heterogeneous Nucleation

During solution crystallization, heterogeneous nucleation is the pervasive process. It is the energy barrier that makes it the preferred nucleation process—heterogeneous nucleation energy barrier is only a fraction of the energy barrier of homogenous nucleation. Two nucleation retardation factors acting simultaneously during heterogeneous nucleation have been anticipated in [16]: (1) occupation of nucleation-active particles and/or centers (generally known as nucleants), associated with the nucleation process itself; and (2) appearance of nucleation exclusion zones (NEZ) formed around growing nuclei. NEZ gradually engulf some of the active nucleants, such that are situated close enough to the formed nuclei, lie in the arising NEZ and are deactivated. This process starts soon after nucleation onset. However, as seen, NEZ do not change the overall system supersaturation.

Accounting for the two retardation factors acting in parallel, and under constant supersaturation, the rate of new-phase heterogeneous nucleation (dn/dt) is expressed by the following single first-order non-linear ordinary differential equation [16]:

$$\frac{dn}{dt} = kn\left(1 - \frac{n}{n_s}\right) \tag{1}$$

Depending on the k-values, nucleation processes can be categorized into the following groups: (1) fast kinetics, e.g., electrochemical nucleation, characterized by very large k-values (of orders 10^3 to 10^4 (s^{-1})); (2) slow kinetics, e.g., protein crystal nucleation (insulin, bovine β-lactoglobulin), k-values of order 10^{-3} to 10^{-4} (s^{-1}); and (3) extremely slow kinetics, e.g., crystallization of cordierite glass, $k \approx 10^{-5}$ (s^{-1}); see in [16].

Integration of Equation (1) gives [16]:

$$n/n_s = 1/[1 + \exp[(-k(t - t_c)] \tag{2}$$

In fact, Equation (1) is a special case of Bernoulli differential equation. Substituting the dimensionless functions $y(x) = n/n_s$ and $x = k(t - t_c)$ in it, and with $t_c = $ const. we have:

$$dy/dx = (n/n_s)[(n_s - n)/n_s] = y(x)[1 - y(x)] \tag{3}$$

The solution of Equation (3) is the standard logistic function:

$$y(x) = \exp(x)/[\exp(x) + C]$$
$$f(x) = e\,x\,e\,x + C\ \{\displaystyle f(x) = \{\frac{e^{x}}{e^{x} + C}\} \tag{4}$$

With constant of integration $C = 1$ $\{\displaystyle C = 1\}$ $C = 1$, this gives the logistic curve definition: $y(x) = \exp(x)/[\exp(x) + 1] = 1/[1 + \exp(-x)]$, which is Equation (2).

Equation (1) shows that the maximum nucleation rate is reached when nucleation acceleration and deceleration tendencies equilibrate, at time t_c, when $n = n_s/2$:

$$(dn/dt)_{max} = kn_s/4 \tag{5}$$

which is the (quasi-)stationary nucleation rate, mentioned above.

Unambiguity of the Logistic Nucleation Time-Dependence

Figures 1 and 2 exemplify the high goodness ($R^2 > 0.99$) of logistic plots. Considered from a physical perspective (as presented above), the good fit of experimental n/n_s vs. t/t_c data for insulin crystal nucleation (Figure 4a,b) shows more stringently the logistic nucleation time-dependence. Firstly, recalling that $n = n_s$ when $t = 2t_c$, $x = k(t - t_c)$ results in $x_{ns} = kt_c = $ const., this explains the self-adjustment between k and t_c occurring for all supersaturations. Secondly, the (orange) logistic curves in Figure 4a,b result from the logistic equation with $2kt_c = 10$ (see the inserts in the figures). Hence, these are standard logistic functional plots with $\pm kt_c = 5$. Due to the function exponential nature, the standard logistic function obtains its real values in the range of $x = \pm 5$ on both sides of its midpoint (Figure 5); in the case under consideration, the latter being at $n_s/2$. It is logical to conclude that an x-value from -6 to -5 can be attributed to the so-called nucleation induction time.

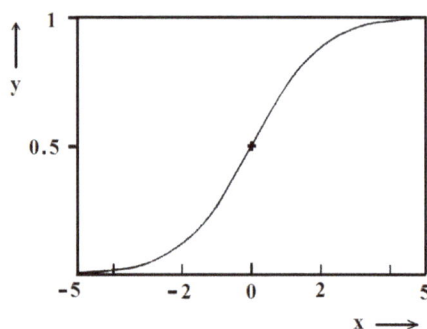

Figure 5. Standard logistic function; (0), (0.5) are the coordinates of the midpoint; in the case considered, it is $n_s/2$.

Further, an almost linear increase of n_s on $\Delta\mu$ is observed for insulin (BioChemika, $\geq$85% (GE), ~24 IU/mg) crystal nucleation in bulk solution, Figure 6. However, it is highly improbable that sets of nucleants possessing nucleation-promoting abilities which correspond exactly to each supersaturation used are present. It is rather a situation where lesser nucleants are engulfed by NEZ (and thus, deactivated) at higher supersaturations.

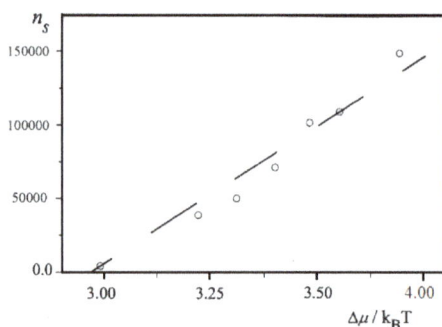

Figure 6. Dependence of n_s in bulk insulin solution vs. dimensionless supersaturation, $\Delta\mu/k_BT = \ln(c/c_e)$.

2.2.2. Rate of Homogeneous Nucleation

Notwithstanding the substantially higher supersaturation required, homogeneous nucleation is indispensable in systems without nucleants. Removal of all nucleants from a protein solution is not an easy task, albeit achievable in the vapor phase. For instance, liquid droplets nucleate homogeneously by rapidly expanding and cooling exceptionally pure water vapors. An exponential increase in water droplet nucleation rate has been measured by means of different techniques [19]. However, homogeneous nucleation could also pose an issue because no unlimited nuclei augmentation is physically feasible. An evident obstacle for observing nucleation rate limits may be the uncountable number of the nucleated droplets. Additional experimental work is needed to compare nucleation rate measurement data to theoretical considerations. Until then, a theoretical approach to the issue is worth attempting.

Like the heterogeneous case, homogeneous nucleation should be a self-limiting process. Again, there are two factors decelerating it. The first one is like the one in heterogeneous nucleation, namely, increase in the number of NEZ appearing around some nuclei and diminishing the volume where nucleation can still occur. The second decelerating factor, namely, a drop in system's overall supersaturation, is different. It is enacted only during prolonged nuclei growth and consumption of a noticeable molecule amount. Furthermore, while the two retardation factors act in parallel in a heterogeneous nucleation, in a homogeneous process they act consecutively, being interrupted by an intermediate period. This constitutes the substantial difference between both processes.

Effect of NEZ

As already mentioned (see Section 2.2), the overall supersaturation remains constant initially. Thus, the probability (n/α) for NEZ appearance, where α (s) is the time needed for the formation of the very first NEZ, also remains constant. The initial nucleation rate $(dn/dt)_{init}$ can be expressed as:

$$(dn/dt)_{init} = K'n \tag{6}$$

where K $(s^{-1}) \neq k$ and $K' = K - 1/\alpha$ are new coefficients.

Preserving the exponential character during the initial nucleation stage, the graphical n vs. t track of the homogeneous nucleation is indistinguishable from the corresponding part of the heterogeneous nucleation curve. However, knowing that merely one decelerating factor is acting, the homogeneous n vs. t dependence should be steeper and relatively longer.

Effect of Decreasing Supersaturation

Increasing in number, soon or latter, the NEZ start overlapping. This indicates that the first nucleation decelerating factor is of no importance on the account of the second one—decrease in system's overall supersaturation. Intermediately, new nuclei appear in the remaining interstitials between NEZ, but there is a substantial deceleration in the exponential increase of n. When $\Delta\mu$ decreases bellow the nucleation-limiting threshold, n vs. t dependence should reach a plateau, corresponding to a zero nucleation rate. Supersaturation dependence of nucleation rate is given by the well-known equation of Volmer. For the second homogeneous nucleation stage, it should be written as:

$$(dn/dt)_{second} = A\exp(-\Delta G^*/k_B T) = A\exp(-B/\Delta\mu^2) \tag{7}$$

where A is a pre-exponential coefficient which denotes the number of nuclei that appear in a unit volume (1 cm^3) per unit time (1 s); ΔG^* is the thermodynamic energy barrier for nucleus formation; the constant B for homogeneously formed spherical nucleus is B = $16\pi\Omega^2\gamma^3/3$ (because ΔG^* = $16\pi\Omega^2\gamma^3/3\Delta\mu^2$); and Ω is the volume of a crystal building block. Qualitatively, this behavior of the system gives a S-shaped dependence of n on time elapsed, t. However, despite the different scenarios involved in heterogeneous and homogeneous nucleation, the exponential functions make their S-shaped time dependent nuclei number densities indistinguishable.

Equation (7) shows that a symmetric S-shape (logistic) curve may describe the homogeneous nucleation, only provided $\Delta\mu$ depends linearly on t, i.e., $\Delta\mu = -st$, where $-s$ is the line slope. Under constant temperature, however, supersaturation decrease results from nuclei growth itself, making the linear dependence physically infeasible. Since the new-phase particles nucleate at different time-points, they grow in different sizes and the size difference is amplified due to the Gibbs-Thomson effect [20]—smaller crystals grow slower than larger crystals; the reason being that the larger the crystal, the lower the saturation with which it stands in equilibrium. That is why, along with an increase in number, nucleated crystals accelerate their growth with the time and increase the rate of supersaturation depletion. In view of the extremely high sensitivity of $(dn/dt)_{second}$ on $\Delta\mu$-value expressed by Equation (7), only the precise function of $\Delta\mu$ on t (but not its linear substitute) is meaningful. Thus, in contrast to the symmetric S-shaped (logistic) curve describing heterogeneous nucleation, a non-symmetric S-shaped curve should describe the n vs. t dependence for homogeneous nucleation.

2.3. Biochemical Specificity of the Protein Crystal Nucleation

In proteins, it is only the molecule surface structure that dictates protein ability to bind to partners. This is attributed to the molecular interactions in protein bulk concealed under amino acid residues situated on the molecular surface. Because of millions of years of natural selection, physiological protein-protein bonds are highly specific. Proteins operate within the cellular context with typical concentrations of up to 300 mg/mL. Therefore, any non-specific inter-protein interaction may be fatal. It is known that physiological protein-protein bonds result from strong hydrophobic interactions via which contacting areas occupy relatively large portions on the protein molecule surface.

In contrast, the protein crystal lattice contacts are hydrophilic, polar and smaller in size [21]. Yet again, it is only the molecular surface structure that dictates proteins' ability to bind to partners in a crystallization setting. In such a setting, a limited number of discrete patches, that are the only attractive molecule portions, appear on the protein surface. If supersaturation is extremely high, amorphous precipitation will occur even under crystallization conditions; such a disordered aggregation is a result of very strong hydrophobic protein-protein interactions. Therefore, it is logical to assume that attraction strength between crystallizing protein molecules should be fine-tuned. Attraction should be large enough to promote crystallization, while not being too large to provoke amorphous precipitation. This means that also protein crystal lattice contacts are formed by a selection of the most appropriate patches on the protein molecule surface. Selection preferences have been revealed using X-ray diffraction data for protein crystal lattice contacts available in Protein Data Bank [21,22].

Strict selection of crystal lattice contacting patches is also evidenced by relatively simple experiments [23]. Periodically alternated layer-by-layer crystal overgrowth has been observed with the unique protein couple apo- and holoferritin. Despite the dramatically different core, their surface structure is identical. Uniform in thickness overlaying crystal layers have been deposited using equimolar protein concentrations under the same solution conditions, pH-value, $CdSO_4$, and buffer concentrations. Since no reentrant corners have been observed (Figure 7), those crystals should be single-crystals composed of alternating apo- and holoferritin layers, rather than poly-crystals. Crystals of each protein are used as substrates for a sequential in contiguity crystallization of the counterpart protein in a completely repeatable process. A monocrystalline overgrowth of three to four alternating layers apo- on holoferritin, and vice versa, was achieved [23]. A clear distinction is allowed as the layers are of different color (apoferritin crystals are yellowish; holoferritin crystals—reddish-brown).

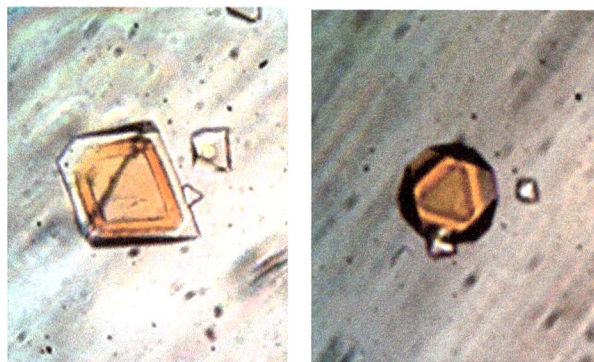

Figure 7. Different orientation of layered monocrystals of apoferritin and holoferritin. Crystal sizes ~200 μm and ~100 μm, respectively.

In contrast, no homoepitaxial monocrystalline overgrowth is possible with proteins possessing differing molecule surfaces. Apoferritin crystals have been purposefully introduced in solutions designed for lysozyme crystallization. No single-crystalline overgrowth, but merely formation of poly-crystalline lysozyme-apoferritin aggregates, has been observed [23]. This shows that a molecule attachment to the protein crystal lattice does not occur at random. It requires selection of the binding partner. It is worth noting that no binding selection is needed for small-molecule crystallization, e.g., by electrodeposition of metal alloys.

Selection of protein-protein patchy interactions has been accounted for by the so-called bond selection mechanism (BSM) [24]. It assumes that a successful collision between protein molecules, leading to formation of a crystalline connection, requires not only sufficiently close proximity of the protein molecules (respectively molecules to clusters), but also their proper spatial orientation. Because relatively small fractions of molecule surface are occupied by contacting patches, the arising steric restriction to protein-protein association postpones the nucleation process significantly. Thus, based on the biochemical specificity of proteins, BSM explains the slow protein crystal nucleation kinetics [25]. Although requiring unusually high supersaturation, it is orders of magnitude slower in comparison to the process with small molecule substances, e.g., during electrochemical nucleation [13,14]. Recalling that crystal nucleation rate changes with process stages, one can only compare the (quasi)-stationary nucleation rates expressed by Equation (5). As seen in Section 2.2.1, k-values determining nucleation frequency are 6 to 8 orders of magnitude lower for protein crystallization than k-values for small molecule new-phase nucleation, e.g., electrochemical nucleation (also proceeding in solutions). So, due to BSM, a much lower attachment frequency ($\nu_R{}^*$) of molecules to the critical cluster must be in place in the pre-exponential coefficient of Volmer's equation:

$$(dn/dt)_{st} = c_1 \nu_R{}^* Z \exp(-\Delta G^*/k_B T) \qquad (8)$$

where c_1 is solute concentration, and Z is known as Zeldovich factor.

3. Materials and Methods

Insulin crystal nucleation kinetics was studied via the so-called nucleation and growth separation principle. Two different insulin sorts, from BioChemika (BioChemika, ≥85% (GE), ~24 IU/mg) and from SIGMA, Denmark, Lot # 080M1589V, were used under identical crystallization conditions. BioChemika-insulin was shown to be more prone to crystal nucleation than SIGMA-insulin. Because more crystals ensure better statistics, BioChemika-insulin was preferred in our studies. Sufficient details allowing replication of the experimental studies are provided in the original paper [17].

4. Conclusions

The early stages of crystal nucleation dictate crystal polymorph selection, which is of great interest to the pharmaceutical industry. Unfortunately, our understanding of these stages remains insufficient [26]. Because of the molecular-scale involved, numerous specifics of nucleation remain largely unknown. Even with state-of-the-art measurements, it is exceptionally challenging to probe the processes in real time. Moreover, new-phase embryos are not labeled, making it impossible to distinguish them in the vast ensemble of constantly growing and decaying clusters of different sizes. The aim of this paper is to shed some additional light on the problem.

A physical aspect of crystal nucleation is considered from the fluctuation-based perspective to cover both CNT and TSNM. Logistic functional dependences according to Equations (1) and (2), symmetric S-shaped curves, characterize the heterogeneous nucleation, while homogeneous nucleation obeys non-symmetric S-shaped functional dependences. Due to the highly inhomogeneous (patchy) surface, proteins are characterized by highly directional interactions which postpone substantially protein crystal nucleation. This is a biochemical constraint imposed on the process. Provided molecule surface patches enabling crystal lattice formation are known, the so-called BSM hypothesis may help in offering clues to proper polymorph selection. Suitable crystal polymorphs can be grown by changing adequately solution conditions (and/or protein molecule surface residues), thus, activating or deactivating different surface patches. However, it is worth also noticing that the precipitants used as crystallizing agents play a specific role [27].

Acknowledgments: The author would like to acknowledge networking support by the COST Action CM1402 "Crystallize". This work is co-financed by a grant from the National Science Fund of the Bulgarian Ministry of Science and Education, contract DCOST 01/22. The experimental help of F. Hodzhaoglu is gratefully acknowledged. The author is grateful to V. Tonchev for drawing the plots.

Conflicts of Interest: The author declares no conflict of interest.

References and Notes

1. Neutron crystallography requires growth of substantially larger protein crystals, greater than 0.1 mm^3 in size are preferred, i.e., 4–5 orders of magnitude larger than those used in synchrotron X-ray data collection [28]. The power of neutron crystallography consists in higher precision by visualization of H-atoms (which play essential roles in macromolecular structure and catalysis), thus helping scientists to understand enzyme reaction mechanisms and hydrogen bonding.

2. Hammadi, Z.; Grossier, R.; Zhang, S.; Ikni, A.; Candoni, N.; Morin, R.; Veesler, S. Localizing and Inducing Primary Nucleation. *Farad. Discuss.* **2015**, *179*, 489–501. [CrossRef] [PubMed]

3. Wölk, J.; Strey, R.; Heath, C.H.; Wyslouzil, B.E. Empirical function for homogeneous water nucleation rates. *J. Chem. Phys.* **2016**, *117*, 4954–4960. [CrossRef]

4. Ten Wolde, P.R.; Frenkel, D. Enhancement of protein crystal nucleation by critical density fluctuations. *Science* **1997**, *277*, 1975–1978. [CrossRef] [PubMed]

5. Vivares, D.; Kaler, E.; Lenhoff, A. Quantitative imaging by confocal scanning fluorescence microscopy of protein crystallization via liquid-liquid phase separation. *Acta Crystallogr. D Biol. Crystallogr.* **2005**, *61*, 819–825. [CrossRef] [PubMed]

6. Sauter, A.; Roosen-Runge, F.; Zhang, F.; Lotze, G.; Jacobs, R.M.J.; Schreiber, F. Real-Time Observation of Nonclassical Protein Crystallization Kinetics. *J. Am. Chem. Soc.* **2015**, *137*, 1485–1491. [CrossRef] [PubMed]

7. Schubert, R.; Meyer, A.; Baitan, D.; Dierks, K.; Perbandt, M.; Betzel, C. Real-Time Observation of Protein Dense Liquid Cluster Evolution during Nucleation in Protein Crystallization. *Cryst. Growth Des.* **2017**, *17*, 954–958. [CrossRef]

8. Sleutel, M.; Van Driessche, A.E.S. Role of clusters in nonclassical nucleation and growth of protein crystals. *Proc. Natl. Acad. Sci. USA* **2014**, *111*, E546–E553. [CrossRef] [PubMed]

9. Sleutel, M.; Lutsko, J.; Van Driessche, A.E.S.; Duran-Olivencia, M.A.; Maes, D. Observing classical nucleation theory at work by monitoring phase transitions with molecular precision. *Nat. Commun.* **2014**, *5*, 5598. [CrossRef] [PubMed]

10. Whitelam, S. Control of Pathways and Yields of Protein Crystallization through the Interplay of Nonspecific and Specific Attractions. *Phys. Rev. Lett.* **2010**, *105*, 088102. [CrossRef] [PubMed]

11. Yamazaki, T.; Kimura, Y.; Vekilov, P.G.; Furukawa, E.; Shirai, M.; Matsumoto, H.; Van Driessche, A.E.S.; Tsukamoto, K. Two types of amorphous protein particles facilitate crystal nucleation. *Proc. Natl. Acad. Sci. USA* **2017**, *114*, 2154–2159. [CrossRef] [PubMed]

12. The patchy character of the protein surfaces is essential for their biological role.

13. Kaischew, R.; Mutaftschiew, B. Ueber die Elektrolytische Keimbildung des Quecksilbers. *Electrochim. Acta* **1965**, *10*, 643–650. [CrossRef]

14. Toschev, S.; Gutzow, I. Nichtstationäre Keimbildung: Theorie und Experiment. *Cryst. Res. Technol.* **1972**, *7*, 43–73. [CrossRef]

15. Interested merely in measuring stationary nucleation rates determined from the slopes of the linear parts of n vs. t dependences at fixed $\Delta\mu$, many researchers disregard the maximum crystal number density (n_s) because sometimes it is difficult to measure it. Complete functional dependences from $n = 0$ till $n = n_s$ are relatively rare, though in virtually all cases they are S-shaped.

16. Nanev, C.N.; Tonchev, V.D. Sigmoid Kinetics of Protein Crystal Nucleation. *J. Cryst. Growth* **2015**, *427*, 48–53. [CrossRef]

17. Nanev, C.N.; Hodzhaoglu, F.V.; Dimitrov, I.L. Kinetics of Insulin Crystal Nucleation, Energy Barrier, and Nucleus Size. *Cryst. Growth Des.* **2011**, *11*, 196–202. [CrossRef]

18. Kertis, F.; Khurshid, S.; Okman, O.; Kysar, J.W.; Govada, L.; Chayen, N.E.; Erlebacher, J. Heterogeneous nucleation of protein crystals using nanoporous gold nucleants. *J. Mater. Chem.* **2012**, *22*, 21928–21934. [CrossRef]

19. Wyslouzil, B.E.; Wölk, J. Overview: Homogeneous nucleation from the vapor phase—The experimental science. *J. Chem. Phys.* **2016**, *145*, 211702. [CrossRef]

20. The newly nucleated crystals are small enough to obey it.

21. Dasgupta, S.; Iyer, G.H.; Bryant, S.H.; Lawrence, C.E.; Bell, J.A. Extent and Nature of Contacts between Protein Molecules in Crystal Lattices and between Subunits in Protein Oligomers. *Proteins Struct. Funct. Genet.* **1997**, *28*, 494–514. [CrossRef]

22. Sergeyev, I.V.; McDermott, A.E. ACCEPT-NMR: A New Tool for the Analysis of Crystal Contacts and Their Links to NMR Chemical Shift Perturbations. *J. Crystalliz. Process Technol.* **2013**, *3*, 12–27. [CrossRef]

23. Nanev, C.N.; Dimitrov, I. Layered crystals of apo- and holoferritin grown by alternating crystallization. *Cryst. Res. Technol.* **2009**, *44*, 908–914. [CrossRef]

24. Nanev, C.N. Kinetics and Intimate Mechanism of Protein Crystal Nucleation. *Prog. Cryst. Growth Charact. Mater.* **2013**, *59*, 133–169. [CrossRef]

25. Alternatively, the latter has been attributed to TSNM [29].

26. Tahri, Y.; Kozisek, Z.; Gagnière, E.; Chabanon, E.; Bounahmidi, T.; Mangin, D. Modeling the competition between polymorphic phases: Highlights on the effect of Ostwald Ripening. *Cryst. Growth Des.* **2016**, *16*, 5689–5697. [CrossRef]

27. Fudo, S.; Qi, F.; Nukaga, M.; Hoshino, T. Influence of Precipitants on Molecular Arrangements and Space Groups of Protein Crystals. *Cryst. Growth Des.* **2017**, *17*, 534–542. [CrossRef]

28. Chen, J.C.-H.; Unkefer, C.J. Fifteen years of the Protein Crystallography Station: The coming of age of macromolecular neutron crystallography. *IUCrJ* **2017**, *4*, 72–86. [CrossRef] [PubMed]

29. Vekilov, P.G. Nucleation of protein crystals. *Prog. Cryst. Growth Charact. Mater.* **2016**, *62*, 136–154. [CrossRef]

![crystals logo]

crystals

MDPI

Review

Biomineralization Mediated by Ureolytic Bacteria Applied to Water Treatment: A Review

Dayana Arias [1,2], Luis A. Cisternas [2,3] and Mariella Rivas [1,3,*]

[1] Laboratory of Algal Biotechnology & Sustainability, Faculty of Marine Sciences and Biological Resources, University of Antofagasta, Antofagasta 1240000, Chile; dayana.arias.t@gmail.com

[2] Department of Chemical Engineering and Mineral Process, University of Antofagasta, Antofagasta 1240000, Chile; luis.cisternas@uantof.cl

[3] Science and Technology Research Center for Mining CICITEM, Antofagasta 1240000, Chile

* Correspondence: mariella.rivas@uantof.cl

Academic Editor: Jolanta Prywer

Received: 6 October 2017; Accepted: 4 November 2017; Published: 17 November 2017

Abstract: The formation of minerals such as calcite and struvite through the hydrolysis of urea catalyzed by ureolytic bacteria is a simple and easy way to control mechanisms, which has been extensively explored with promising applications in various areas such as the improvement of cement and sandy materials. This review presents the detailed mechanism of the biominerals production by ureolytic bacteria and its applications to the wastewater, groundwater and seawater treatment. In addition, an interesting application is the use of these ureolytic bacteria in the removal of heavy metals and rare earths from groundwater, the removal of calcium and recovery of phosphate from wastewater, and its potential use as a tool for partial biodesalination of seawater and saline aquifers. Finally, we discuss the benefits of using biomineralization processes in water treatment as well as the challenges to be solved in order to reach a successful commercialization of this technology.

Keywords: biomineralization; calcite; seawater; wastewater; heavy metals removal; biodesalination

---

1. Introduction

Without water, there is no life and, in recent years, this resource is increasingly scarce. Various factors such as climate change, droughts and population increase contribute to its scarcity [1]. Although the earth's surface consists of 70% water, only 3% corresponds to freshwater. In addition, a significant part of that percentage is found in ice caps and glaciers, consequently only 1% of the surface freshwater is usable. Efficiency in water use can significantly increase if the capacity to reuse it increases by using new and improved technologies. In the case of seawater, desalination emerges as an alternative to extract salt and other polluting elements from it, turning it into water suitable for human consumption, or for productive uses such as agriculture and mining, among others. Desalination is identified as a safe source of water that guarantees a stable supply as compared with the variability of natural sources and the scarcity of this resource in the basins. However, the application of chemical methods generates waste that affects the environment [2,3].

In contrast, the use of new biotechnological tools could favor the recycle of wastewater and improve the quality of seawater and freshwater. Such is the case of biomineralization, a process that is mediated by bacteria and other organisms for the formation of minerals from ions present in the surrounding environment. A particular form of biomineralization is microbiological carbonates precipitation (MICP), and is defined as the process involving the formation of minerals mediated by living organisms as a result of the cellular activity that promotes the physicochemical conditions required to carry out the formation and growth of the biominerals [4]. There is a great variety of structures and, in nature, more than 60 types of biological minerals have been described [5]. From an evolutionary point of view, this process is generated mainly from bacterial activity. Bacteria are capable of inducing mineral precipitation through three types of mechanisms [6].

i) Biologically controlled mineralization consists of cellular activities specifically aimed at the formation of minerals [6–8]. Organisms direct the synthesis of minerals in a specific part of the cell but only under certain conditions. For example, the formation of magnetite by magnetotactic bacteria. The magnetotactic bacteria have control over the mineral phase and its biosynthesis in the magnetosome, at the level of genes.

ii) Biologically influenced mineralization corresponds to the passive precipitation of minerals due to the interaction between the environment and its chemical and compound changes present in the cellular surface as a result of the bacterial metabolic activity, for example extracellular polymeric substances associated with biofilms [7–10]. In this type of biomineralization, the biominerals are secreted due to the metabolism of the microorganisms, and the system has very little control over the minerals that have been deposited. There are a large number of bacteria capable of biologically inducing the extracellular precipitation of a wide range of minerals, involving the geochemical activity responsible for mineral deposits in terrestrial evolution.

iii) Biologically induced mineralization, which corresponds to the chemical modification of an environment mediated by the biological activity producing oversaturation and precipitation of minerals [6,8,11].

In the third case, the bacteria contribute actively to the formation of minerals, but also in a passive way, through cellular structures that act as nucleation sites [12–15]. In addition, bacteria can induce heterogeneous nucleation of minerals, given not only by a surface with a lower energetic barrier to minerals precipitation, but also with a stereochemical structuring of the mineral components. However, there may be a combination of the three different processes active in the same system [8]. Some studies highlight the role of the extracellular polymer substances (EPS). These exopolymers and the biofilms are commonly dominated by negatively charged polysaccharides [16] and may contribute to the precipitation of minerals in different forms;

i) By trapping positively charged cations in negatively charged sites of the EPS that act as a tempering for crystal nucleation [9,17].

ii) By entrapment of crystal seeds that act as a nuclei for the heterogeneous precipitation of minerals [9,14].

Within this point, biological processes that increase the pH of the medium and create the oversaturated conditions necessary for its precipitation have been identified [4,18]. Different bacteria with varied metabolisms are able to precipitate biominerals from carbonates [19–23], oxides [24,25], sulfates [26,27] and phosphates [28,29]. This review focuses on biomineralization mediated by ureolytic bacteria. The best-studied mechanism is the precipitation of calcium carbonate by ureolysis, in which the bacteria metabolize urea through an intracellular urease enzyme, producing HCO_3^- and NH_3. The latter is converted into NH_4^+, alkalinizing the medium, and the HCO_3^- is converted into CO_3^{2-} [30,31]. When any calcium ion is present and oversaturation of calcite occurs, precipitation of calcium carbonate is induced [32].

2. Ureolytic Metabolism

The urease enzyme is present in a great diversity of microorganisms (urea amidohydrolase; EC 3.5.1.5) reviewed in [6,8,33] enabling the cell to use urea as a source of nitrogen [34]. *Sporosarcina pasteurii* ATCC11859, formally *Bacillus pasteurii* [35], is the terrestrial ureolytic bacterium most used as an example of MICP and presents an active intracellular urease [4,11].

The water treatment processes mediated by ureolytic bacteria described in this review are based on the general mechanism of carbonates biological precipitation, which basically consists of the microorganisms´ ability to alkalize the surrounding environment according to the physiological activities they perform. These bacteria are widely distributed in nature and their role is to catalyze the hydrolysis of urea to produce carbonic acid and ammonium [8,30,36]. These products, in solution, have as final result to induce a change of pH in the medium (Ecs. 1–4) [30]:

$$CO(NH_2)_2 + H_2O \xrightarrow{\text{urease}} NH_3 + CO(NH_2)OH \tag{1}$$

$$CO(NH_2)OH + H_2O \rightarrow NH_3 + H_2CO_3 \tag{2}$$

$$H_2CO_3 \leftrightarrow HCO_3^- + H^+ \tag{3}$$

$$2NH_3 + 2H_2O \rightarrow 2NH_4^+ + 2OH^- \tag{4}$$

The increase in pH leads to an adjustment of the bicarbonate equilibrium to form carbonate ions, further favoring the formation of CO_3^{2-} from HCO_3^- [37]. A high carbonate concentration induces $CaCO_3$ precipitation around the cells and the presence of calcium ions in the surrounding environment (Ecs. 5–7) [38].

$$2HCO_3^- + 2H^+ + 2NH_3 + 2OH^- \leftrightarrow 2CO_3^{2-} + 2NH_4^+ + 2H_2O \tag{5}$$

$$Ca^{2+} + Cell \rightarrow Cell : Ca^{2+} \tag{6}$$

$$Cell : Ca^{2+} + CO_3^{2-} \rightarrow Cell : CaCO_3 \tag{7}$$

Under natural conditions, the precipitation of carbonates occurs very slowly. In this sense, microorganisms would act as catalysts in the carbonate formation process. Carbonates, especially calcite ($CaCO_3$) and dolomite ($CaMg(CO_3)_2$) are found as limestones on the Earth's surface, representing an important carbon stock in the lithosphere [38,39].

The mechanisms through which the biological precipitation of carbonates occurs are not fully described. However, three mechanisms have been proposed to explain this process [40]: i) Biomineralization occurs as a bioproduct of the microbial metabolism; ii) Extracellular molecules are involved in the carbonate mineralization process; iii) A nucleation process of carbonates occurs in the cell wall of microorganisms. Based on the latter proposed mechanism, the role of microorganisms in creating an alkaline environment through various physiological activities is known. Under these circumstances, the bacterial surface plays an important role in the precipitation of carbonates due to the presence of various negatively charged groups at neutral pH, positive ions can bind to the bacterial surface favoring heterogeneous nucleation [40]. The microbiological precipitation of minerals has several technological applications, such as the restoration of limestone monuments and statues, biocement production, improvement of soil quality and removal of soluble pollutants such as heavy metals and radioactive elements [30]. Additionally, this process promoted specifically by the ureolytic capacity of bacterial species also allows the removal of secondary ions such as calcium and magnesium present in wastewater [33,41,42] and in seawater [43]. Furthermore, the precipitation of calcium carbonates mediated by ureolytic bacteria is widely described in the literature, mainly under the application of soil biocementation [8,44,45]. Other uses are related to carbon dioxide capture and remediation of soils and water [40,46].

3. Types of Biominerals Produced by Ureolytic Bacteria

Compared with inorganically produced minerals, biominerals often have their own specific properties including unique size, crystallinity, isotopic and trace element compositions [47]. There are several types of biominerals such as organic crystals, oxides, hydroxides, sulfates, sulfides, chlorides, phosphates and carbonates [39,48–50] highlighting calcium carbonates. Calcium carbonate exists in three different crystalline structures: vaterite, calcite and aragonite, and in two hydrated crystalline phases: monohydrocalcite ($CaCO_3 \cdot H_2O$) and ikaite ($CaCO_3 \cdot 6H_2O$), and various amorphous phases (ACC) [8,40]. Although the crystalline structures of carbonates due to bacterial activity are clear, little is known about the causes of the selection of different polymorphisms during bacterial biomineralization.

Although there are studies suggesting that the amount and morphology of calcium carbonate minerals depend on oversaturation, temperature, pH, $[Ca^{2+}]/[CO_3{}^{2-}]$ ratio [51] and the concentration of urea [42]. Hammes et al. [33] during a study of ureolytic microbial calcium carbonate ($CaCO_3$) precipitation by bacterial isolates collected from different environmental samples, morphological differences were observed in the large $CaCO_3$ crystal aggregates precipitated within bacterial colonies grown on agar. This fact was verified at our laboratory confirming that different halophilic and halotolerant ureolytic bacterial strains isolated from the Atacama Salar produce calcium carbonate crystals with different polymorphism when they were cultivated in nutrient broth (NB) supplemented with Ca^{2+} (Figure 1). This is exactly what Uad et al. [50] described: biominerals present a variety of forms from spheres, hemispheres and pseudopolyhedral forms, which appeared either in isolation or in group.

Although calcium carbonates are the most studied crystals, ureolytic bacteria are capable of forming different types of crystals depending on the medium in which they are (Table 1). For example, vaterite is found to be in a lower percentage, metastable or in the transitional phase of calcite [8] and it has been described that it depends on the concentration of EPS and the organic matter that would influence its formation, which could be stabilized in the presence of certain organics [8,52]. The maturation of $CaCO_3$ from vaterite to calcite follows the Ostwald's step rule, where metastable forms nucleate and then are replaced with more stable forms, with this sequential formation in time is also known as paragenesis [53]. However, the mechanisms of initial nucleation that are influenced by bacterial growth conditions, the presence of organic matter such as EPS, saturation conditions of the fluid and crystals maturation are still not well understood [8,54]. The size of crystals is also a key factor. The $CaCO_3$ crystals precipitated by ureolytic metabolism are generally large and less soluble than those precipitated under abiotic conditions [51,55].

Table 1. Examples of crystals produced by ureolytic bacteria.

Bacterial Species	Crystal	Aplication	Reference
Enterobacter cloacae	Calcite	Heavy metals bioremediation	Kang et al [56]
Bacillus sp.	Calcitemagnesium carbonate trihydrate ($MgCO_3·3H_2O$)	Biocementation	Cheng et al. [57]
Sporosarcina ginsengisoli	As(III)–calcite calcite, aragonite and vaterite	As(III) remediation	Achal et al. [58]
Halomonas sp.	calcite, vaterite and aragonite along with calcite–strontianite ($SrCO_3$) solid Halite (NaCl)	Sr remediation	Achal et al. [59]
Rhodococcus erythropolis	Monohydrocalcite ($CaCO_3·H_2O$) Struvite Anhydrite	Calcium and magnesium precipitation from sea water	Arias et al. [43]
Strains of Bacillus sphaericus group	rhombohedral calcite, hexagonal vaterite		Hammes et al. [33]

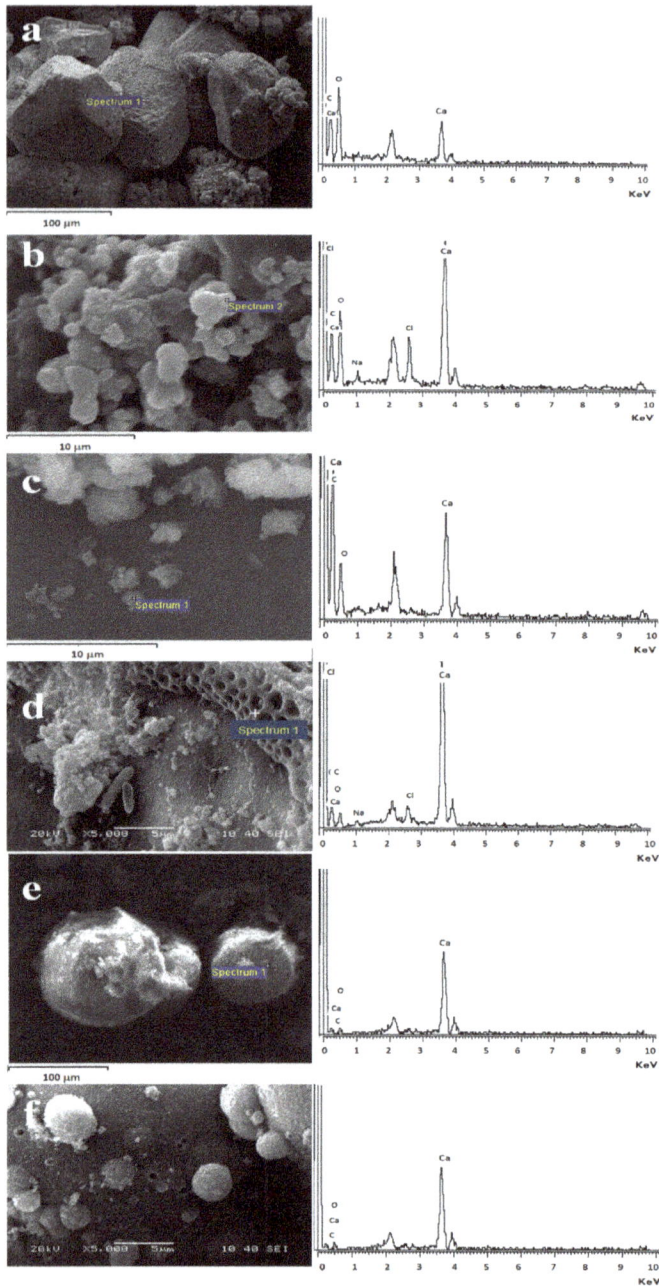

Figure 1. Scanning electron microscopy with energy dispersive X-ray (SEM-EDX) analysis of calcium carbonate crystals produced by various bacterial ureolytic species in aqueous solution supplemented with Ca^{2+}. Crystals produced by *B. subtilis* LN8B (**a**), *Halomonas* sp. LM12ABN (**b**), *Rhodococcus erythropolis* TN24F (**c**), *Bacillus subtilis* TN21G (**d**), *Salinivibrio* sp. LM9A (**e**) and *Bacillus subtilis* LN13DC (**f**). Methodology can be reviewed in [43].

The use of ions present in seawater for crystal formation mediated by ureolytic bacteria, specifically *Bacillus* sp. DSM23526 strain MCP-11 [60] was first described by Cheng et al. [57]. *Bacillus* sp. induced calcite crystals formation, additionally the presence of magnesium carbonate trihydrate ($MgCO_3 \cdot 3H_2O$) was detected, which is a consequence of the high concentration of magnesium ion in seawater (above 50 mM), which is five times more than the calcium concentration (10 mM) [57].

From calcite precipitation induced by ureolytic bacteria, some authors have evaluated to add divalent metals (e.g., Pb, Zn, Ba and Cd) and radionuclides (e.g., ^{90}Sr and ^{60}Co) [61] in a coprecipitation of the contaminants in the original calcite precipitates, which would occur by the isomorphic replacement of the Ca^{2+} in the lattice structure and the incorporation in the interstitial positions or at defect vacancies [62,63] generating an important application for biomineralization as a remediation strategy for contaminated groundwater [63].

4. Application of Ureolytic Bacteria for the Removal of Heavy Metals and Radionuclides from Aqueous Solutions

The increase in industrial activity has resulted in the accumulation of a number of pollutants such as heavy metals and radionuclides that cause problems both to the environment and to human health. These come from industrial effluents, mining activities, waste from the use of fertilizers and pesticides in agriculture, burning of waste and fossil fuels, and leaching of dumps [47]. Urban solid waste dumps are the common method for the organized disposal of waste in the world and the produced leachate contains a wide variety of organic and inorganic pollutants that occasionally can reach groundwater.

The most important pollutants include metals such as Cu, Cr, Cd, Hg, Sb, Pb, As, Co, Zn and Sn, and radionuclides such as U, Th, Am, Ra and Sr [38]. The latter is one of the most studied through the use of ureolytic bacteria, which comes from various sources such as plastics, concrete, plasters or other manufactured products [64]. The danger with heavy metals and radionuclides is that they are not chemically or biologically degradable and once used and/or released, they can remain in the environment for hundreds of years, and their concentration in living beings increases as they are ingested by the members of the food chain [65].

Regarding heavy metals, there are several conventional techniques for their elimination including chemical precipitation, oxidation/reduction, filtration, ion exchange, reverse osmosis, membrane technology, evaporation method and electrochemical treatment. However, these traditional methods often are ineffective, expensive, consume energy and high quantities of chemicals [6]. When the concentration of the heavy metals is less than 100 mg/L, these technologies become inefficient leading to the use of bioremediation technologies [6,66].

One of the most used bioremediation technologies in the removal of heavy metals mainly from soils is phytoremediation. However, it has some limitations due to reliance on plant growth conditions such as climate, geology, altitude and temperature [58].

Alternatively, there are other soil and water remediation technologies based on the use of microorganisms mediated by different bioremediation mechanisms, including biosorption, metal–microbe interactions, bioaccumulation, biotransformation, bioleaching and finally, biomineralization [66].

So far, the most described biomineralization technology is mediated by the action of ureolytic microorganisms. One of the advantages of using this type of microorganism for bioremediation of metals from soil or water is its ability to efficiently immobilize these toxic metals by precipitation or coprecipitation with $CaCO_3$, dependent of metal valence status and redox potential [38]. In addition, its high stability and the coprecipitation potential of radionuclides or heavy metals is an attractive application of the ureolytic bacteria use [6].

Ammonium ions can exchange heavy metal ions and other ions like calcium on grain surfaces in subsurface environment, and this in turn increases the bioavailability of these heavy metals [67]. On the other hand, the carbonate ions promote the precipitation of calcium carbonate and coprecipitation of heavy metals in high pH environments [68,69]. Although most of the studies are related to the bioremediation of contaminated soils, they begin with tests on aqueous solutions and the immobilization process may enable metal(loid)s to be transformed in situ into insoluble and chemically inert forms and are applicable to removing metals from aqueous solution. For example, Achal et al. [70] describe the ability of *Kocuria flava* CR1 to remove Cu from the environment at concentrations of 100 mg/L. In the presence of urea, *K. flava* is able to remove 95% of copper in 120 h, without urea it removes only 68% evidencing the ureolytic character of the metabolism used by the bacterium; this is corroborated with an optimum pH of 8 for a complete removal of copper. Another study by Achal et al. [58] assesses tolerance to As(III) of *Sporosarcina ginsengisoli* CR5 isolated from a site contaminated with As in Urumqi, China. Bioassays in the presence of urea and 50 mM of As(III) indicated that this strain is capable of synthesizing ureases for 7 days of a test, with a maximum production of 412U/mL at 120 h, remediating 96.3% of the As present. The introduction of this indigenous bacterium provides a potential bioremediation process to highly metal-contaminated water and soils.

Among the examples, Isik [71] uses ureolytic bacteria as nickel adsorbents by comparing a living and nonliving ureolytic mixed culture (UMC) to remove Ni^{2+} from synthetic wastewater solutions, demonstrating a greater efficiency in the living culture dependent on ureolytic metabolism. For chromium, a study by Altaş et al. [72] determined the potential as sorbent for Cr^{6+} from an aqueous solution and regarding to Cu^{2+}, Simsek et al. [73] confirmed the ability of UMC for its adsorption by ureolytic bacteria. This last study was carried out under *batch* conditions, removing 99% Cu^{2+} from an initial solution with 100 mg/L of this metal.

Several studies have also been carried out regarding the removal of radionuclides, particularly Mitchell and Ferris [63] analyzed the coprecipitation of Sr^{2+}, establishing nucleation and growth of calcite precipitates by bacterial ureolysis in artificial groundwater free of Sr replicating the composition of a contaminated aquifer. It was established that in the presence of Sr^{2+}, the calcite crystals had an average diameter of <840 nm as compared, the diameter in the absence of Sr^{2+} had a gradual size increase of <1000 nm. The speed of growth of the crystals is limited by the speed of advection of the solute to the surface of the crystal. The crystals generated in the presence of Sr are smaller and, therefore, more soluble. However, it does not significantly reduce the long-term effectiveness of Sr immobilization, obtaining 99% of the calcite precipitation and the coprecipitate of Sr.

Another example is described by Achal et al. [59], who determined the precipitation of calcite by ureolysis to remediate the radioactive Sr (^{90}Sr) present in the quartz sand of an aquifer using the *Halomonas* sp. bacterium, resistant to this element. It was determined an 80% removal of ^{90}Sr from the soluble–exchangeable fraction of the quartz sand of the aquifer. X-ray diffraction detected calcite, vaterite and aragonite along with a solid solution of calcite-strontianite ($SrCO_3$) in a bioremediated sample indicating that Sr was incorporated into calcite. This demonstrates that through biomineralization, the soluble Sr in the form of biomineral is abducted and playing an important role in the bioremediation of Sr from the ecological and economic point of view. The Sr has also been removed using the *Bacillus pasteurii* bacterium [69] from synthetic waters whose composition simulates an aquifer from the Snake River plain, Idaho. It was possible to demonstrate the co-precipitation of calcite with Sr by X-ray diffraction and Time-of-flight secondary ion mass spectrometry (ToF SIMS), proving that Sr was not only absorbed at the surface, but was present in depth within the particles. This is determined by the global rate of calcite precipitation, while the latter indicates a higher absorption of Sr in the solid. Subsequently, Fujita et al. [68] confirmed the presence of the *ureC* gene in waters and sediments in Sr removal tests in situ with combined calcite precipitation.

5. Phosphorus Precipitation from Wastewaters

The high demographic and industrial growth in recent years have led to an increase in water pollution. In particular, nutrient discharges into natural waters have contributed to an increase in eutrophication problems, resulting in serious consequences for aquatic life as well as for the provision of water for industrial and domestic use [74]. In the case of the food industry related to the production of frozen vegetables and prefried potatoes, the main problem lies in the presence of large amounts of phosphate in wastewater, leading to its accumulation in the process water unless elimination of phosphate occurs before to be reused [75]. One of the major contaminants due to this process is struvite. Struvite ($NH_4MgPO_4 \cdot 6H_2O$) is the most common form of magnesium phosphate found in nature [76]. Regarding wastewater treatment plants, the areas most affected by struvite deposition are places where there is an increase in turbulence, such as pumps, aerators and pipe curves [77].

Although struvite is a problem in wastewater treatment plants, it also has a potential use as fertilizer [78]. Moreover, other benefits of struvite precipitation include the reduction of phosphorus and nitrogen loading on secondary mud and recycled mud to the top of the list of wastewater treatment works [79]. Phosphorus recovery not only prevents eutrophication but also preserves limited natural resources.

The conventional method to eliminate phosphorus from wastewater is the addition of flocculants such as ferric chloride and aluminum [80]. However, the main problem is the high economic cost of these chemical flocculants, and in many cases, this process can be replaced by the biological crystallization or biomineralization of phosphorus [80]. There are three types of commercial technologies for phosphorus biological disposal, called Phostrip, modified processes of Bardenpho and Rotanox [81].

Several studies have been published that focus on the treatment of wastewater both to prevent eutrophication and to recover phosphorus and its potential use as a fertilizer using the ureolytic metabolism of bacteria. Carballa et al. [78] describe a method to remove phosphate from wastewater from anaerobic effluents by ureolytic precipitation. This study applied phosphate precipitation mediated by ureolytic bacteria when the concentration of this ion in wastewater from a vegetable processing industry was below <30 mg PO_4^{3-} -P levels by liter. MgO (preferably $Mg(OH)_2$) and urea were added to facilitate precipitation, decreasing its concentration to approximately 5−7 mg PO_4^{3-}-P per liter. A study by Desmidt et al. [75] evaluated the removal of phosphate from industrial anaerobic effluents. They first modeled the precipitation of magnesium ammonium phosphate (MAP) as struvite with MAPLE and PHREEQC programs; however, the obtained experimental results were better than those expected with precipitation models. According to this, they indicated that a pH above 8.5 and the use of $MgCl_2$ produced better results compared with the use of MgO for the phosphate precipitation from industrial anaerobic effluent. Subsequently, Desmidt et al. [82] used water from a potato processing company for the comparison of MAP precipitation using ureolytic crystallization combined with an autotrophic nitrogen removal process with NuReSys®technology, which increases pH by NaOH and recovers orthophosphate (PO_4−P) and ammoniacal nitrogen (NH_4-N) to form pure struvite crystals ($NH_4MgPO_4 \cdot 6H_2O$) (http://www.nuresys.be/). The importance of this comparison is to establish that MAP ureolytic crystallization is competitive with NuReSys®technology in terms of operational cost and removal efficiency; however, further research is still needed to obtain more crystals. Desmidt et al. [83] determined that several factors influence struvite precipitation, including calcium concentration, concentration of PO_4−P, Mg^{2+}:PO_4^{3-} molar ratio and ionic strength. Despite good results, the fact that the wastewater contains a low concentration of PO_4−P influences the application of this technology economically since the addition of other chemicals like urea to carry out the process is necessary.

Another application in the recovery of phosphorus (P) by ureolytic bacteria corresponds to its removal from urine using seawater. Dai et al. [84] determined that 98% of P from urine can be precipitated in the presence of seawater, in only 10 minutes when 40–75% of the urea present in the urine is hydrolyzed, while magnesium and ammonium phosphate (MAP) was the predominant component of the precipitates found. However, in this study they do not mention ureolytic bacteria unlike Tang et al. [85], who describe the Seawater-based Urine Phosphorus Recovery (SUPR) system indicating that native urine and/or seawater bacteria provide a ureolytic population in situ facilitating phosphate precipitation in SUPR.

6. Precipitation of Ions from Wastewater and Seawater

Calcium present in wastewater and seawater generates several operational problems—mainly scaling in pipelines and reactors due to its precipitation as sulphate, phosphate and/or carbonate salt [86]. Regarding seawater, this has a typical salinity of 35 g/kg of solution and the NaCl concentration around 0.5–0.6 M. In addition to this salt, seawater contains a number of important secondary ions. Most cations such as calcium, magnesium and strontium are free; however, there is a tendency to form ionic pairs with sulphate [87]. To remove calcium, there are chemical crystallization reactors based on the addition of a base (NaOH or $Ca(OH)_2$) in the presence of nucleation sites for the precipitation of crystals; however, most of the time these are expensive, complex and sometimes generate highly alkaline effluents due to the reagents used.

For example, in the mining industry, the seawater is used without desalination, affecting mineral flotation of copper-molybdenum due to the presence of high concentrations of secondary ions, such as Mg^{2+} and Ca^{2+}, which generate a buffer effect [86,88] and increases the consumption of reagents used to modify the pH in the solution. In the case of adding lime or another alkalizing agent, colloidal precipitates of these ions are produced which prevent the recovery of some relevant species such as molybdenite (MoS_2) [88–91]. In addition, the Cl^- ion can react with the Fe^{2+} ions present to create $FeCl_2$. This compound reacts with the dissolved oxygen producing Fe_2O_3 and $FeCl_3$, which are considered strong oxidizing agents of pipes and/or equipment [91]. Another mineral originated through this mechanism is calcium carbonate ($CaCO_3$), the most common sedimentary mineral present on Earth [30].

The application of biomineralization processes mediated by ureolytic bacteria for calcium removal and other ions from seawater and wastewater is poorly described.

One of the earliest studies is by Hammes et al. [33], which looked at the removal of calcium by biomineralization from industrial wastewater (Ca^{2+} 500–1500 mg/L) through the use of a semicontinuous reactor (BCR), removing about 90% of calcium in the period under test. The water came from a paper recycling facility. Although they do not specifically detail the type of microorganisms used, they attribute calcium precipitation to ureolytic metabolism. They also suggest that the chemical precipitation of calcium carbonate is controlled by four factors: i) calcium ion concentration; ii) carbonate concentration; iii) pH and, iv) presence of nucleation sites. In the microbial precipitation of calcium carbonate, the first three factors are key, while the fourth is replaced by the presence of bacteria as they act as nucleation sites themselves. Then Işik et al. [42] determined the effect of urea concentration on the BCR reactor yield in terms of the removal of organic matter over time and, secondly, evaluated the ammonia toxicity in the ureolytic mixed culture of S. *pasteurii* confirming that this bacterium isolated from alkaline environments (pH values of 9–13) can induce the calcite biomineralization. Furthermore, this bacterium is widely used in controlled laboratory conditions and in the presence of industrial solid waste such as cement kiln dust (CKD) and lime kiln dust (LKD). Its potential for grain consolidation for a type of CKD mixed with granulated blast furnace slag (GGBS), has potential application in bioclogging and biocementation. The results demonstrate the formation of stable biocalcite in the presence of CKD, with a yield that depends on the pH value and the content of free calcium ions, the benefits of this technology in construction costs and reduction of environmental contamination are promising [92].

In a study developed by Uad et al. [50], the biomineralization of calcium carbonate was carried out by bacteria of the *Bacillus* and *Virgibacillus* genus isolated from saline environments and cultivated in seawater and brines from desalination plants, confirming that these species are able to precipitate calcium carbonate when grown in culture media supplemented with organic matter.

Regarding magnesium, most publications on its precipitation indicate that they are part of struvite crystals and are related to the recovery of phosphates from wastewater through chemical crystallization [93]. However, there are also records that biologically mediated struvite precipitation is possible [94]. Various bacteria isolated from waters and soils such as *Myxococcus* sp., *Arthrobacter* sp. and *Pseudomonas* sp. have been described as capable of precipitating struvite under laboratory conditions [94]. Among them, *Myxococcus xanthus* biomineralizes struvite [95] and on the other hand, *Halorubrum distributum*, *Halobacterium salinarium* and *Brevibacterium antiquum* have also been reported as forming magnesium phosphate crystals [96]. However, precipitation of magnesium from seawater using ureolytic bacteria has been poorly described and is related to what has already been described for the precipitation of struvite from wastewaters. Nevertheless, Arias et al. [43] determined that ureolytic bacteria isolated from the Atacama Desert act as biomineralizers of calcium and magnesium ions present in seawater; this could be used as a potential pretreatment technology to seawater or as a selective biodesalation. In this case *Rhodococcus erythropolis* precipitates to ~95% soluble calcium and 8% magnesium. The analysis of crystals showed that the components correspond to ~12.69% monohydrocalcite, ~30.72% struvite and ~56.59% halite, concluding that these results have great potential for their application as a pretreatment to improve water quality for industrial processes.

7. Current Limitations for the Application of Biomineralization

Biomineralization as technology has many interesting applications that could solve many problems without a high economic or environmental cost; however, it is necessary to solve some limitations to reach a commercial scale. Firstly, a complete life cycle analysis of the biomineralization technology should be carried out. In the case of using ureolytic bacteria, as byproducts of the metabolism, ammonium and nitrate are produced, which could be toxic and harmful to human health in high concentration [6,97]. Secondly, microbial metabolic processes are generally slower and more complex than chemical processes; consequently, it is necessary to optimize all factors involved, such as temperature, pH, urea concentration, salts present, considering without doubt the bacterial species to use and the necessary conditions to favor its metabolism. Thirdly, the implementation of technologies based on biomineralization for water treatment requires the expertise of many disciplines including the efforts of engineers, microbiologists, biochemists, among others. Finally, it is necessary to consider the economic limitations, especially in the replacement of reagents and nutrients of analytical grade used in most studies by others of lower cost. Additionally, although some studies, such as those by Hammes et al. [33] and Işik et al. [41], use a semicontinuous reactor, there are no other described technologies that allow for further thought in regards to scale thus far.

8. Conclusions

Biomineralization or MICP provides the basis for numerous technologies that allow for the removal of ions from different types of water including freshwater, wastewater and seawater, which would allow for the discharge and/or a harmless use of these waters. Numerous studies have employed the use of microorganisms whose ureolytic metabolism facilitates the formation of various types of biominerals, with calcite formation the most common. The various applications of ureolytic metabolism include the removal of heavy metals and radionuclides—such as Cu, As, Ni, Cr and Sr, among others—through calcite coprecipitation demonstrating the great potential of this technology in incorporating these elements into low-solubility biominerals that are stable over time and inert compared to phytoremediation technologies. In addition to the above, MICP also facilitates the removal of specific ions from wastewater and seawater, including phosphorus, magnesium and calcium. Even though the MICP process has many merits, further study is needed to overcome the limitations to use this technology prior to its commercialization.

Acknowledgments: This publication was supported by Anillo Programme–Grant n° ACT1201-Atacama Seawater, PhD Scholarship CONICYT n° 21130712, CICITEM Project n° R10C1004 and the Regional Government of Antofagasta.

Author Contributions: DA, LC, and MR contributed equally to the writing of this review.

Conflicts of Interest: The authors declare no conflicts of interest.

References

1. Anaya, M.; Martínez, J.M. Manual de Captación de Agua de Lluvia Para Áreas Rurales: Sistemas de Captación Y Aprovechamiento del Agua de Lluvia Para Uso Doméstico y Consumo Humano en América Latina y el Caribe. México, Instituto de Enseñanza e Investigación en Ciencias Agrícolas. 2007. Available online: http://www.pnuma.org/recnat/esp/public.php (accessed on 15 April 2016).

2. Lattemann, S.; Höpner, T. Environmental Impact and Impact Assessment of Seawater Desalination. *Desalination* **2008**, *220*, 1–15. [CrossRef]

3. Dawoud, M.A.; Al Mulla, M.M. Environmental Impacts of Seawater Desalination: Arabian Gulf Case Study. *Int. J. Environ. Sustain.* **2012**, *1*, 22–37. [CrossRef]

4. DeJong, J.T.; Soga, K.; Banwart, S.; Whalley, W.R.; Ginn, T.R.; Nelson, D.C.; Mortensen, B.M.; Martinez, B.C.; Barkouki, T. Soil Engineering In Vivo: Harnessing Natural Biogeochemical Systems for Sustainable, Multi-functional Engineering Solutions. *J. R. Soc. Interface* **2011**, *8*, 1–15. [CrossRef] [PubMed]

5. Sarikaya, M. Biomimetics: Materials Fabrication through Biology. *Proc. Natl. Acad. Sci. USA* **1999**, *96*, 14183–14185. [CrossRef] [PubMed]

6. Anbu, P.; Kang, C.H.; Shin, Y.-J.; So, J.-S. Formations of Calcium Carbonate Minerals by Bacteria and its Multiple Applications. *SpringerPlus* **2016**, *5*, 250. [CrossRef] [PubMed]

7. Benzerara, K.; Miot, J.; Morin, G.; Ona-Nguema, G.; Skouri-Panet, F.; Férard, C. Significance, Mechanisms and Environmental Implications of Microbial Biomineralization. *C. R. Geosci.* **2011**, *343*, 160–167. [CrossRef]

8. Phillips, A.J.; Gerlach, R.; Lauchnor, E.; Mitchell, A.C.; Cunningham, A.B.; Spangler, L. Engineered Applications of Ureolytic Biomineralization: A Review. *Biofouling* **2013**, *29*, 715–733. [CrossRef] [PubMed]

9. Jimenez-Lopez, C.; Jroundi, F.; Rodriguez-Gallego, M.; Arias, J.M.; González-Muñoz, M.T. Biomineralization induced by *Myxobacteria*. *Commun. Curr. Res. Educ. Top. Trends Appl. Microbiol.* **2007**, *1*, 143–154.

10. Sarayu, K.; Iyer, N.G.; Ramachandra-Murty, A. Exploration on the Biotechnological Aspect of the Ureolytic Bacteria for the Production of the Cementitious Materials—A Review. *Appl. Biochem. Biotechnol.* **2014**, *172*, 2308–2323. [CrossRef] [PubMed]

11. Stocks-Fischer, S.; Galinat, J.K.; Bang, S.S. Microbiological Precipitation of CaCO$_3$. *Soil Biol. Biochem.* **1999**, *31*, 1563–1571. [CrossRef]

12. Chafetz, H.S.; Buczynski, C. Bacterially Induced Lithification of Microbial Mats. *Palaios* **1992**, *7*, 277–293. [CrossRef]

13. Ben Omar, N.; Martínez-Cañamero, M.; González-Muñoz, M.T.; Arias, J.M.; Huertas, F. Struvite Crystallization on *Myxococcus* cells. *Chemosphere* **1995**, *30*, 2387–2396. [CrossRef]

14. Von Knorre, H.; Krumbein, W.E. *Microbial Sediments*; Riding, R., Awaramik, S., Eds.; Springer: Berlin, Germany, 2000; pp. 25–31.

15. Paerl, H.W.; Stepp, T.F.; Reid, R.P. Bacterially Mediated Precipitation in Marine Stromatolites. *Environ. Microbiol.* **2001**, *3*, 123–130. [CrossRef] [PubMed]

16. Decho, A.W. Microbial Biofilms in Intertidal Systems: An Overview. *Cont. Shelf. Res.* **2000**, *20*, 1257–1273. [CrossRef]

17. Pentecost, A. Association of Cyanobacteria with Tufa Deposits: Identity, Enumeration and Nature of the Sheath Material Revealed by Histochemistry. *Geomicrobiol. J.* **1985**, *4*, 285–298. [CrossRef]

18. Whiffin, V.S.; Van Paassen, L.A.; Harkes, M.P. Microbial Carbonate Precipitation as a Soil Improvement Technique. *Geomicrobiol. J.* **2007**, *24*, 417–423. [CrossRef]

19. Wright, D.T. The Role of Sulphate-Reducing Bacteria and Cyanobacteria in Dolomite Formation in Distal Ephemeral Lakes of the Coorong region, South Australia. *Sediment. Geol.* **1999**, *126*, 147–157. [CrossRef]

20. Castanier, S.; Le Métayer-Levrel, G.; Orial, G.; Loubière, J.-F.; Pethuisot, J.P. *Of Microbes and Art: The Role of Microbial Communities in the Degradation and Protection of Cultural Heritage*; Ciferri, O., Ed.; Plenun: New York, NY, USA, 2000; pp. 203–218.

21. Ben Chekroun, K.; Rodriguez-Navarro, C.; Gonzalez-Muñoz, M.T.; Arias, J.M.; Cultrone, G.; Rodriguez-Gallego, M. Precipitation and Growth Morphology of Calcium Carbonate induced by *Myxococcus xanthus*: Implications for Regognition of Bacterial Carbonates. *J. Sediment. Res.* **2004**, *74*, 868–876. [CrossRef]

22. Rivadeneyra, M.A.; Párraga, J.; Delgado, R.; Ramos-Cormenzana, A.; Delgado, G. Biomineralization of Carbonates by *Halobacillus trueperi* in Solid and Liquid Media with Different Salinities. *FEMS Microbiol. Ecol.* **2004**, *48*, 39–46. [CrossRef] [PubMed]

23. Baskar, S.; Baskar, R.; Mauclaire, L.; McKenzie, J.A. Microbially induced Calcite Precipitation by Culture Experiments—Possible Origin for Stalactites in Sahastradhara, Dehradun, India. *Curr. Sci.* **2006**, *90*, 58–64.

24. Coates, J.D.; Anderson, R.T.; Woodward, J.C.; Phillips, E.J.P.; Lovley, D.R. Anaerobic Hydrocarbon Degradation in Petroleum-Contaminated Harbor Sediment Under Sulfate-Reducing and Artificially Imposed Iron-Reducing Conditions. *Environ. Sci. Technol.* **1996**, *30*, 2784–2789. [CrossRef]

25. Ringelberg, D.B.; White, D.C.; Nishijima, M.; Sano, H.; Burghardt, J.; Stackebrandt, E.; Nealson, K.H. Polyphasic taxonomy of the genus Shewanella and description of Shewanella oneidensis sp. nov. *Int. J. Syst. Bacteriol.* **1999**, *49*, 705–724.

26. Jerden, J.L.; Sinha, A.K. Phosphate Based Immobilization of Uranium in an Oxidizing Bedrock Aquifer. *Appl. Geochem.* **2003**, *18*, 823–843. [CrossRef]

27. Jroundi, F.; Merroun, M.; Arias, J.M.; Rossberg, A.; Selenska-Pobell, S.; Gonzalez-Muñoz, M.T. Spectroscopic and Microscopic Characterization of Uranium Biomineralization by *Myxococcus xanthus. Geomicrobiol. J.* **2007**, *24*, 441–449. [CrossRef]

28. Takazoe, I.; Vogel, J.; Ennever, J. Calcium Hydroxyapatite Nucleation by Lipid Extract of *Bacterionema matruchotii. J. Dent. Res.* **1970**, *49*, 395–398. [CrossRef] [PubMed]

29. Smith, W.; Streckfu, J.; Vogel, J.; Ennever, J. Struvite Crystals in Colonies of *Bacterionema matruchotii* and its Variants. *J. Dent. Res.* **1971**, *50*, 777–781. [CrossRef] [PubMed]

30. Al-Thawadi, S.M. Ureolytic Bacteria and Calcium Carbonate Formation as a Mechanism of Strength Enhancement of Sand. *J. Adv. Sci. Eng. Res.* **2011**, *1*, 98–114.

31. Abo-El-Enein, S.A.; Ali, A.H.; Talkhan, F.N.; Abdel-Gawwad, H.A. Utilization of Microbial Induced Calcite Precipitation for Sand Consolidation and Mortar Crack Remediation. *HBRC J.* **2012**, *8*, 185–192. [CrossRef]

32. Bosak, T. Calcite Precipitation, Microbially Induced. In *Encyclopedia of Earth Sciences Series*; Reitner, J., Thiel, V., Eds.; Springer: Dordrecht, The Netherlands, 2012; pp. 223–227.

33. Hammes, F.; Seka, A.; De Knijf, S.; Verstraete, W. A Novel Approach to Calcium Removal from Calcium-Rich Industrial Wastewater. *Water Res.* **2003**, *37*, 699–704. [CrossRef]

34. Burbank, M.B.; Weaver, T.J.; Williams, B.C.; Crawford, R.L. Urease Activity of Ureolytic Bacteria Isolated from Six Soils in which Calcite was Precipitated by Indigenous Bacteria. *Geomicrobiol. J.* **2012**, *29*, 389–395. [CrossRef]

35. Yoon, J.H.; Lee, K.C.; Weiss, N.; Kho, Y.H.; Kang, K.H.; Park, Y.H. *Sporosarcina aquimarina* sp. nov., a Bacterium Isolated from Seawater in Korea, and transfer of *Bacillus globisporus* (Larkin and Stokes 1967), *Bacillus psychrophilus* (Nakamura 1984) and *Bacillus pasteurii* (Chester 1898) to the genus *Sporosarcina* as *Sporosarcina globispora* comb. nov., *Sporosarcina psychrophila* comb. nov. and *Sporosarcina pasteurii* comb. nov., and Emended Description of the Genus *Sporosarcina. Int. J. Syst. Evol. Microbiol.* **2001**, *51*, 1079–1086. [PubMed]

36. Krajewska, B. Ureases I. Functional, Catalytic and Kinetic Properties: A Review. *J. Mol. Catal. B Enzym.* **2009**, *59*, 9–21. [CrossRef]

37. Knoll, A.H. Biomineralization and Evolutionary History. *Rev. Mineral. Geochem.* **2003**, *54*, 329–356. [CrossRef]

38. Kumari, D.; Qian, X.Y.; Pan, X.; Achal, V.; Li, Q.; Gadd, G.M. Microbially-induced Carbonate Precipitation for Immobilization of Toxic Metals. *Adv. Appl. Microbiol.* **2016**, *94*, 79–108. [PubMed]

39. Ehrlich, L.; Newman, D.K. *Geomicrobiology*, 5th ed.; CRC Press; Taylor & Francis Group: New York, NY, USA, 2009.

40. Dhami, N.K.; Reddy, M.S.; Mukherjee, A. Biomineralization of Calcium Carbonates and their Engineered Applications: A Review. *Front. Microbiol.* **2013**, *4*, 314. [CrossRef] [PubMed]

41. Işık, M.; Altaş, L.; Kurmaç, Y.; Özcan, S.; Oruç, Ö. Effect of Hydraulic Retention Time on Continuous Biocatalytic Calcification Reactor. *J. Hazard. Mater.* **2010**, *182*, 503–506. [CrossRef] [PubMed]

42. Işık, M.; Altaş, L.; Özcan, S.; Şimşek, I.; Ağdağ, O.N.; Alaş, A. Effect of Urea Concentration on Microbial Ca Precipitation. *J. Ind. Eng. Chem.* **2012**, *18*, 1908–1911. [CrossRef]

43. Arias, D.; Cisternas, L.A.; Rivas, M. Biomineralization of Calcium and Magnesium Crystals from Seawater by Halotolerant Bacteria Isolated from Atacama Salar (Chile). *Desalination* **2017**, *405*, 1–9. [CrossRef]

44. Dick, J.; De Windt, W.; De Graef, B.; Saveyn, H.; Van der Meeren, P.; De Belie, N.; Verstraete, W. Bio-deposition of a Calcium Carbonate Layer on Degraded Limestone by *Bacillus* species. *Biodegradation* **2006**, *17*, 357–367. [CrossRef] [PubMed]

45. Silva-Castro, G.A.; Uad, I.; Gonzalez-Martinez, A.; Rivadeneyra, A.; Gonzalez-Lopez, J.; Rivadeneyra, M.A. Bioprecipitation of Calcium Carbonate Crystals by Bacteria Isolated from Saline Environments Grown in Culture Media Amended with Seawater and Real Brine. *BioMed Res. Int.* **2015**, *2015*, 816102. [CrossRef] [PubMed]

46. Achal, V.; Mukherjee, A.; Kumari, D.; Zhang, Q. Biomineralization for Sustainable Construction—A review of Processes and Applications. *Earth-Sci. Rev.* **2015**, *148*, 1–17. [CrossRef]

47. Li, M.; Cheng, X.; Guo, H. Heavy Metal Removal by Biomineralization of Urease Producing Bacteria Isolated from Soil. *Int. Biodeterior. Biodegrad.* **2012**, *76*, 81–85. [CrossRef]

48. Barabesi, C.; Galizzi, A.; Mastromei, G.; Rossi, M.; Tamburini, E.; Perito, B. *Bacillus subtilis* Gene Cluster Involved in Calcium Carbonate Biomineralization. *J. Bacterial.* **2007**, *189*, 228–235. [CrossRef] [PubMed]

49. Achal, V.; Mukherjee, A.; Basu, P.C.; Reddy, M.S. Strain Improvement of *Sporosarcina pasteurii* for Enhanced Urease and Calcite Production. *J. Ind. Microbial. Biotechnol.* **2009**, *36*, 981–988. [CrossRef] [PubMed]

50. Uad, I.; Gonzalez-Lopez, J.; Silva-Castro, G.A.; Vílchez, J.I.; Gonzalez-Martinez, A.; Martin-Ramos, D.; Rivadeneyra, M.A. Precipitation of Carbonates Crystals by Bacteria Isolated from a Submerged fixed-film Bioreactor used for the Treatment of urban Wastewater. *Int. J. Environ. Res.* **2014**, *8*, 435–446.

51. Mitchell, A.C.; Ferris, F.G. The Influence of *Bacillus pasteurii* on the Nucleation and Growth of Calcium Carbonate. *Geomicrobiol. J.* **2006**, *23*, 213–226. [CrossRef]

52. Rodriguez-Navarro, C.; Jimenez-Lopez, C.; Rodriguez-Navarro, A.; Gonzalez-Muñoz, M.T.; Rodriguez-Gallego, M. Bacterially mediated Mineralization of Vaterite. *Geochim Cosmochim. Acta* **2007**, *71*, 1197–1213. [CrossRef]

53. Morse, J.W.; Casey, W.H. Ostwald Processes and Mineral Paragenesis in Sediments. *Am. J. Sci.* **1988**, *288*, 537–560. [CrossRef]

54. Zamarreño, D.; Inkpen, R.; May, E. Carbonate Crystals precipitated by Freshwater Bacteria and their use as a Limestone Consolidant. *Appl. Environ. Microbiol.* **2009**, *75*, 5981–5990. [CrossRef] [PubMed]

55. Mitchell, A.C.; Phillips, A.J.; Schultz, L.; Parks, S.; Spangler, L.; Cunningham, A.; Gerlach, R. Microbial CaCO$_3$ Mineral Formation and Stability in an Experimentally Simulated High Pressure Saline Aquifer with Supercritical CO$_2$. *Int. J. Greenh. Gas Control* **2013**, *15*, 86–96. [CrossRef]

56. Kang, C.-H.; Kwon, Y.-J.; So, J.-S. Bioremediation of heavy metals by using bacterial mixtures. *Ecol. Eng.* **2016**, *89*, 64–69. [CrossRef]

57. Cheng, L.; Ha Hin, M.A.S.; Cord-Ruwisch, R. Bio-cementation of Sandy Soil using Microbially induced Carbonate Precipitation for Marine Environments. *Géotechnique* **2014**, *64*, 1010–1013. [CrossRef]

58. Achal, V.; Pan, X.; Fu, Q.; Zhang, D. Biomineralization Based Remediation of As (III) Contaminated Soil by *Sporosarcina ginsengisoli*. *J. Hazard. Mater.* **2012**, *201*, 178–184. [CrossRef] [PubMed]

59. Achal, V.; Pan, X.; Zhang, D. Bioremediation of Strontium (Sr) Contaminated Aquifer Quartz Sand based on Carbonate Precipitation induced by Sr Resistant *Halomonas* sp. *Chemosphere* **2012**, *89*, 764–768. [CrossRef] [PubMed]

60. Al-Thawadi, S.M.; Cord-Ruwisch, R.; Bououdina, M. Consolidation of Sand Particles by Nanoparticles of Calcite after Concentrating Ureolytic Bacteria in situ. *Int. J. Green Nanotechnol.* **2012**, *4*, 1–9. [CrossRef]

61. Knobel, L.L.; Bartholomay, R.C.; Cecil, L.D.; Tucker, B.J.; Wegner, S.J. *Chemical Constituents in the Dissolved and Suspended Fractions of Groundwater from Selected Sites, Idaho National Engineering Laboratory and Vicinity, Idaho, 1989*; Open-File Report Geological Survey: Idaho Falls, ID, USA, 1992.

62. Reeder, J.R.; Lamble, G.M.; Northrup, P.A. XAFS Study of the coordination and Local Relaxation around Co^{2+}, Zn^{2+}, Pb^{2+} and Ba^{2+} trace elements in Calcite. *Am. Mineral.* **1999**, *84*, 1049–1060. [CrossRef]

63. Mitchell, A.C.; Ferris, F.G. The Coprecipitation of Sr into Calcite precipitates induced by bacterial Ureolysis in Artificial Groundwater: Temperature and Kinetic Dependence. *Geochim. Cosmochim. Acta* **2005**, *69*, 4199–4210. [CrossRef]

64. Nigro, A.; Sappa, G.; Barbieri, M. Strontium Isotope as Tracers of Groundwater Contamination. *Procedia Earth Planet. Sci.* **2017**, *17*, 352–355. [CrossRef]

65. He, J.; Chen, J.P. A Comprehensive Review on Biosorption of Heavy Metals by Algal Biomass: Materials, Performances, Chemistry, and Modelling Simulation Tools. *Bioresour. Technol.* **2014**, *160*, 67–78. [CrossRef] [PubMed]

66. Dixit, R.; Malaviya, D.; Pandiyan, K.; Singh, U.B.; Sahu, A.; Shukla, R.; Paul, D. Bioremediation of Heavy Metals from Soil and Aquatic Environment: an Overview of Principles and Criteria of Fundamental Processes. *Sustainability* **2015**, *7*, 2189–2212. [CrossRef]

67. Wu, Y.J.; Ajo-Franklin, J.B.; Spycher, N.; Hubbard, S.; Zhang, G.; Williams, K.; Taylor, J.; Fujita, Y.; Smith, R. Geophysical Monitoring and Reactive Transport Modeling of Ureolytically driven Calcium Carbonate Precipitation. *Geochem. Trans.* **2011**, *12*, 7. [CrossRef] [PubMed]

68. Fujita, Y.; Taylor, J.L.; Wendt, L.M.; Reed, D.W.; Smith, R.W. Evaluating the Potential of Native Ureolytic Microbes to Remediate a ^{90}Sr contaminated Environment. *Environ. Sci. Technol.* **2010**, *44*, 7652–7658. [CrossRef] [PubMed]

69. Fujita, Y.; Redden, G.D.; Ingram, J.C.; Cortez, M.M.; Ferris, F.G.; Smith, R.W. Strontium Incorporation into Calcite Generated by Bacterial Ureolysis. *Geochim. Cosmochim. Acta* **2004**, *68*, 3261–3270. [CrossRef]

70. Achal, V.; Pan, X.; Zhang, D. Remediation of Copper-Contaminated Soil by *Kocuria flava* CR1, based on Microbially induced Calcite Precipitation. *Ecol. Eng.* **2011**, *37*, 1601–1605. [CrossRef]

71. Işik, M. Biosorption of Ni(II) from Aqueous Solutions by Living and Non-living Ureolytic Mixed Culture. *Colloids Surf. B Biointerfaces* **2008**, *62*, 97–104. [CrossRef] [PubMed]

72. Altaş, L.; Kiliç, A.; Koçyiĝit, H.; Işik, M. Adsorption of Cr(VI) on Ureolytic Mixed culture from Biocatalytic Calcification Reactor. *Colloids Surf. B Biointerfaces* **2011**, *86*, 404–408. [CrossRef] [PubMed]

73. Simsek, I.; Karatas, M.; Basturk, E. Cu(II) Removal from Aqueous Solution by Ureolytic mixed Culture (UMC). *Colloids Surf. B Biointerfaces* **2013**, *102*, 479–483. [CrossRef] [PubMed]

74. Pastor, L.; Mangin, D.; Barat, R.; Seco, A. A Pilot-scale Study of Struvite Precipitation in a Stirred Tank Reactor: Conditions Influencing the Process. *Bioresour. Technol.* **2008**, *99*, 6285–6291. [CrossRef] [PubMed]

75. Desmidt, E.; Verstraete, W.; Dick, J.; Meesschaert, B.D.; Carballa, M. Ureolytic Phosphate Precipitation from Anaerobic Effluents. *Water Sci. Technol.* **2009**, *59*, 1983–1988. [CrossRef] [PubMed]

76. Rivadeneyra, M.A.; Pérez-García, I.; Ramos-Cormenzana, A. Influence of Ammonium Ion on bacterial Struvite Production. *Geomicrobiol. J.* **1992**, *10*, 125–137. [CrossRef]

77. Stratful, I.; Scrimshaw, M.D.; Lester, J.N. Conditions Influencing the Precipitation of Magnesium Ammonium Phosphate. *Water Res.* **2001**, *35*, 4191–4199. [CrossRef]

78. Carballa, M.; Moerman, W.; De Windt, W.; Grootaerd, H.; Verstraete, W. Strategies to Optimize Phosphate Removal from Industrial Anaerobic Effluents by Magnesium Ammonium Phosphate (MAP) Production. *J. Chem. Technol. Biotechnol.* **2009**, *84*, 63–68. [CrossRef]

79. Doyle, J.D.; Parsons, S.A. Struvite Formation, Control and Recovery. *Water Res.* **2002**, *36*, 3925–3940. [CrossRef]

80. Alamdari, A.; Rohani, S. Phosphate Recovery from Municipal Wastewater through Crystallization of Calcium Phosphate. *Pac. J. Sci. Technol.* **2007**, *8*, 27–31.

81. Stratful, I.; Brett, S.; Scrimshaw, M.B.; Lester, J.N. Biological Phosphorus Removal, its Role in Phosphorous Recycling. *Environ. Technol.* **1999**, *20*, 681–695. [CrossRef]

82. Desmidt, E.; Ghyselbrecht, K.; Monballiu, A.; Verstraete, W.; Meesschaert, B.D. Evaluation and Thermodynamic Calculation of Ureolytic Magnesium Ammonium Phosphate Precipitation from UASB Effluent at Pilot Scale. *Water Sci. Technol.* **2012**, *65*, 1954–1962. [CrossRef] [PubMed]

83. Desmidt, E.; Ghyselbrecht, K.; Monballiu, A.; Rabaey, K.; Verstraete, W.; Meesschaert, B.D. Factors influencing Urease Driven Struvite Precipitation. *Sep. Purif. Technol.* **2013**, *110*, 150–157. [CrossRef]

84. Dai, J.; Tang, W.T.; Zheng, Y.S.; Mackey, H.R.; Chui, H.K.; van Loosdrecht, M.C.; Chen, G.H. An Exploratory Study on Seawater-catalysed Urine Phosphorus Recovery (SUPR). *Water Res.* **2014**, *66*, 75–84. [CrossRef] [PubMed]

85. Tang, W.; Dai, J.; Liu, R.; Chen, G. Microbial Ureolysis in the Seawater-catalysed Urine Phosphorus Recovery System: Kinetic Study and Reactor Verification. *Water Res.* **2015**, *87*, 10–19. [CrossRef] [PubMed]

86. Castro, S. Flotación con Agua de mar. In *El Agua de Mar en la Minería: Fundamentos y Aplicaciones*; Cisternas, L., Moreno, L., Eds.; RIL Editores: Santiago, Chile, 2014; pp. 99–119. ISBN 978-956-01-0081-8.

87. Castro, S.; Laskowski, J.S. Froth Flotation in Saline Water. *KONA Powder Part. J.* **2011**, *29*, 4–15. [CrossRef]

88. Castro, S. Challenges in Flotation of Cu-Mo Sulfide Ores in Sea Water. In *Proceedings of the 1st International Symposium, Water in Mineral Processing*; Drelich, J., Ed.; SME: Englewood, IL, USA, 2012; pp. 29–40.

89. Castro, S.; Lopez-Valdivieso, A.; Laskowski, J.S. Review of the Flotation of Molybdenite. Part I: Surface Properties and Floatability. *Int. J. Miner. Process* **2016**, *148*, 48–58. [CrossRef]

90. Castro, S.; Laskowski, J.S. Depressing Effect of Flocculants on Molybdenite Flotation. *Miner. Eng.* **2015**, *74*, 13–19. [CrossRef]

91. Liang, J.; Deng, A.; Xie, R.; Adin, A. Impact of Seawater Reverse Osmosis (SWRO) Product Remineralization on the Corrosion Rate of Water Distribution Pipeline Materials. *Desalination* **2013**, *311*, 54–61. [CrossRef]

92. Cuzman, O.A.; Rescic, S.; Richter, K.; Wittig, L.; Tiano, P. *Sporosarcina pasteurii* use in Extreme Alkaline Conditions for Recycling Solid Industrial Wastes. *J. Biotechnol.* **2015**, *214*, 49–56. [CrossRef] [PubMed]

93. Le Corre, K.S.; Valsami-Jones, E.; Hobbs, P.; Parsons, S.A. Phosphorus Recovery from Wastewater Struvite Crystallization: A Review. *Crit. Rev. Environ. Sci. Technol.* **2009**, *39*, 433–477. [CrossRef]

94. Pratt, C.; Parsons, S.A.; Soares, A.; Martin, B.D. Biologically and Chemically mediated Adsorption and Precipitation of Phosphorus from Wastewater. *Curr. Opin. Biotechnol.* **2012**, *23*, 890–896. [CrossRef] [PubMed]

95. Gonzalez-Muñoz, M.T.; Rodriguez-Navarro, C.; Martínez-Ruiz, F.; Arias, J.M.; Merroun, M.L.; Rodriguez-Gallego, M. Bacterial Biomineralization: New Insights from *Myxococcus*-induced Mineral Precipitation. *Geol. Soc. Lond. Spec. Publ.* **2010**, *336*, 31–50. [CrossRef]

96. Smirnov, A.; Suzina, N.; Chudinova, N.; Kulakovskaya, T.; Kulaev, I. Formation of Insoluble Magnesium Phosphates During Growth of the Archaea *Halorubrum* distributum and *Halobacterium salinarium* and the *Brevibacterium antiquum*. *FEMS Microbiol. Ecol.* **2005**, *52*, 129–137. [CrossRef] [PubMed]

97. Van Paassen, L.A.; Daza, C.M.; Staal, M.; Sorokin, D.Y.; van der Zon, W.; van Loosdrecht, M.C.M. Potential Soil Reinforcement by Biological Denitrification. *Ecol. Eng.* **2010**, *36*, 168–175. [CrossRef]

MDPI

St. Alban-Anlage 66

4052 Basel

Switzerland

Tel. +41 61 683 77 34

Fax +41 61 302 89 18

www.mdpi.com

Crystals Editorial Office

E-mail: crystals@mdpi.com

www.mdpi.com/journal/crystals

www.ingramcontent.com/pod-product-compliance
Lightning Source LLC
Chambersburg PA
CBHW051915210326

41597CB00033B/6151